ENVIRONMENTAL REMEDIATION TECHNOLOGIES, REGULATIONS AND SAFETY

ENVIRONMENTAL PROTECTION AGENCY AND TOXIC CHEMICALS MANAGEMENT

ENVIRONMENTAL REMEDIATION TECHNOLOGIES, REGULATIONS AND SAFETY

Additional books in this series can be found on Nova's website under the Series tab.

Additional E-books in this series can be found on Nova's website under the E-book tab.

ENVIRONMENTAL REMEDIATION TECHNOLOGIES, REGULATIONS AND SAFETY

ENVIRONMENTAL PROTECTION AGENCY AND TOXIC CHEMICALS MANAGEMENT

ROBIN LAURITS
EDITOR

nova publishers
New York

For permission to use material from this book please contact us:
Telephone 631-231-7269; Fax 631-231-8175
Web Site: http://www.novapublishers.com

Library of Congress Cataloging-in-Publication Data

ISBN: 978-1-62808-244-9

Published by Nova Science Publishers, Inc. † New York

CONTENTS

PREFACE

Tens of thousands of chemicals are listed with the Environmental Protection Agency (EPA) for commercial use in the United States, with an average of 600 new chemicals listed each year. The EPA's ability to effectively implement its mission of protecting public health and the environment depends on credible and timely assessments of the risks posed by toxic chemicals. In 1976, Congress passed the Toxic Substances Control Act (TSCA) to provide the EPA with the authority to obtain more information on chemicals and to regulate those chemicals that the EPA determines pose unreasonable risks to human health or the environment. The TSCA authorizes the EPA to review chemicals already in commerce and chemicals yet to enter commerce. In 2009, the EPA announced TSCA reform principles to inform ongoing efforts to manage toxic chemicals. This book assesses the increased efforts of the EPA to ensure the safety of toxic chemicals management.

Chapter 1 – In 1976, Congress passed TSCA to provide EPA with the authority to obtain more information on chemicals and to regulate those chemicals that EPA determines pose unreasonable risks of injury to human health or the environment. GAO has reported that EPA has found much of TSCA difficult to implement—hampering the agency's ability to obtain certain chemical data or place limits on chemicals. Of the thousands of chemicals listed for commercial use in the United States, EPA has used its authority to limit or ban five chemicals since TSCA was enacted. In 2009, EPA announced TSCA reform principles to inform ongoing efforts in Congress to strengthen the act. At that time, EPA also initiated a new approach for managing toxic chemicals with the goal of ensuring the safety of chemicals using its existing authorities.

GAO was asked to evaluate EPA's efforts to strengthen its management of chemicals. This report determines the extent to which (1) EPA has made progress implementing its new approach and (2) EPA's new approach positions it to achieve its goal of ensuring the safety of chemicals. GAO examined agency documents and TSCA rulemaking and interviewed agency officials and stakeholders from industry and environmental organizations.

Chapter 2 – This report summarizes the Toxic Substances Control Act (TSCA) and the major regulatory programs dealing with chemical production and distribution in U.S. commerce.

The President's Council on Environmental Quality proposed comprehensive federal legislation in 1971 to identify and control potentially dangerous chemicals in U.S. commerce that were not adequately regulated under other environmental statutes. President Ford signed TSCA into law on October 11, 1976. Subsequently, five titles were added to address specific concerns—asbestos in 1986 (Title II, P.L. 99-519), radon in 1988 (Title III, P.L. 100-551), lead in 1992 (Title IV, P.L. 102-550), environmental and energy issues in schools in 2007 (Title V, P.L. 110-140), and formaldehyde emissions from composite wood products in 2010 (Title VI, P.L. 111-199). In 2008, Congress added provisions to Title I, Section 6 and Section 12, banning certain activities with respect to elemental mercury (P.L. 110-414).

TSCA authorizes EPA to identify potentially dangerous chemicals in U.S. commerce that should be subject to federal control. The act authorizes EPA to gather and disseminate information about production, use, and possible adverse effects to human health and the environment of existing chemicals, and to issue "test rules" that require manufacturers and processors of potentially dangerous chemicals to conduct and report the results of scientific studies to fill information gaps. For chemicals new to U.S. commerce, TSCA requires pre-market screening and regulatory tracking of new chemical products.

If EPA identifies unreasonable risks associated with existing or new chemicals, TSCA requires the agency to initiate rulemaking to reduce risks to a reasonable level. EPA may regulate the manufacture, importation, processing, distribution, use, and/or disposal of chemicals. TSCA provides a variety of regulatory tools to EPA, ranging in severity from a total ban on production, import, and use to a requirement that a product must bear a warning label at the point of sale. However, TSCA directs EPA to use the least burdensome option that can reduce risk to a level that is reasonable, given the benefits provided by the chemical product or process.

Title I of the original statute establishes the core program, directs EPA to control risks from polychlorinated biphenyls (PCBs), and bans certain activities with respect to elemental mercury. Title II directs EPA to set standards for asbestos mitigation in schools and requires asbestos contractors to be trained and certified. Title III directs EPA to provide technical assistance to states that choose to support radon monitoring and control. Title IV provides similar assistance with respect to abatement of lead-based paint hazards. Title V addresses environmental issues at schools, including energy efficiency. Finally, Title VI directs EPA to set standards for emissions of formaldehyde from composite wood products.

Chapter 3 – The U.S. Environmental Protection Agency (EPA) is committed to working with the Congress, members of the public, the environmental community, and the chemical industry to reauthorize the Toxic Substances Control Act (TSCA). The Administration believes it is important to work together to quickly modernize and strengthen the tools available in TSCA to increase confidence that chemicals used in commerce, which are vital to our Nation's economy, are safe and do not endanger the public health and welfare of consumers, workers, and especially sensitive sub-populations such as children, or the environment.

The following Essential Principles for Reform of Chemicals Management Legislation (Principles) are provided to help inform efforts underway in this Congress to reauthorize and significantly strengthen the effectiveness of TSCA. These Principles present Administration goals for updated legislation that will give EPA the mechanisms and authorities to expeditiously target chemicals of concern and promptly assess and regulate new and existing chemicals.

Chapter 4 – The TSCA inventory of chemicals in commerce now exceeds 84,000 chemicals. Periodic TSCA chemical data reporting indicates that there are approximately 7,000 chemicals currently produced at volumes of 25,000 pounds or greater. Under TSCA, EPA is charged with the responsibility of assessing the safety of these commercial chemicals and to act upon those chemicals if there are significant risks to human health or the environment. A sustained and predictable approach is needed to effectively carry out this responsibility.

EPA believes that this significant and long-term challenge can best be met via legislative reform of TSCA to improve EPA's chemical management authorities. Until reform is achieved, however, EPA's responsibility to create a sustained and effective existing chemicals program must be carried out under current authorities.

This strategy summarizes EPA's approach in 2012 and beyond, pending legislative reform.

Given the vast number of chemicals; the high cost to EPA of performing comprehensive risk assessments, and, if appropriate, risk management; and the Agency's responsibility to protect human health and the environment, EPA has developed the following multi-pronged approach for the Agency's existing chemicals management program:

1. Risk assessment and risk reduction
2. Data collection and screening
3. Public access to chemical data and information

In summary, the Agency will perform risk assessments and, if appropriate, risk management for those chemicals with well-characterized hazard concerns and which present the possibility of significant exposure. These are likely to be a relatively small number of chemicals, compared to the size of the universe of commercial chemicals. While risk assessments are being conducted for this small group of chemicals, EPA will be developing an approach to screen the thousands of other compounds to determine which ones warrant further attention, which could include comprehensive risk assessments, or additional data development, addressing either hazard or exposure. Many chemicals will likely be judged as being of lower concern. Finally, EPA will work toward making chemical information available. In particular, the Agency will work to ensure that hazard and exposure data are available to the public in a manner that is most useful to those in the public who will be using the information. Taking this approach to address multiple aspects of the chemicals management challenge simultaneously should allow the Agency to be more comprehensive in its efforts, despite the large number of high production chemicals.

Chapter 5 – In the Agency's August 2011 *Discussion Guide: Background and Discussion Questions for Identifying Priority Chemicals for Review and Assessment*, EPA described the two-step process the Agency intended to use to identify potential candidate chemicals for near-term review and assessment under the Toxic Substances Control Act (TSCA). The Agency intends to use these TSCA Work Plan Chemicals to help focus and direct the activities of the Existing Chemicals Program in the Office of Pollution Prevention and Toxics (OPPT). *EPA invited public comment* through an *online discussion forum* conducted from August 18 through September 21, 2011, as well as through a webinar and stakeholder meeting held on September 7, 2011. The meeting summaries and public comments are available for review in the docket for this activity, EPA-HQ-OPPT-2011-0516, which can be accessed online at http://www.regulations.gov.

As described in the *Discussion Guide*, EPA notes that identification of a chemical as a TSCA Work Plan Chemical does not itself constitute a finding by the Agency that the chemical presents a risk to human health or the environment. Such a determination would be the result of a risk assessment. Rather, identification of a chemical as a TSCA Work Plan Chemical indicates only that the Agency intends to consider it for further review. The Agency believes that identifying these chemicals early in the review process would afford all interested parties the opportunity to bring additional relevant information on those chemicals to the Agency's attention in order to further inform the review. In order to take risk management actions on a chemical substance under various sections of TSCA, the Agency would have to make the appropriate findings required by the specific provisions of the statute.

Identification of some chemicals as TSCA Work Plan Chemicals (Work Plan) does not mean that EPA would not consider other chemicals for risk assessment and potential risk management action under TSCA and other statutes. EPA will consider other chemicals if warranted by available information. In addition, EPA may subsequently identify other candidates for review in addition to this initial group, and may adapt the factors and data sources used in this process based on the experience acquired during this initial phase. Further, while the chemicals identified through this process as TSCA Work Plan Chemicals will likely be well-characterized for hazard and have information indicating exposure potential, some will have more limited data and EPA will continue to use its TSCA information collection, testing, and subpoena authorities, including sections 4, 8, and 11(c) of TSCA, to develop needed information on additional chemicals that currently have less robust hazard or exposure databases.

In: Environmental Protection Agency ... ISBN: 978-1-62808-244-9
Editor: Robin Laurits

Chapter 1

TOXIC SUBSTANCES: EPA HAS INCREASED EFFORTS TO ASSESS AND CONTROL CHEMICALS BUT COULD STRENGTHEN ITS APPROACH*

United States Government Accountability Office

WHY GAO DID THIS STUDY

In 1976, Congress passed TSCA to provide EPA with the authority to obtain more information on chemicals and to regulate those chemicals that EPA determines pose unreasonable risks of injury to human health or the environment. GAO has reported that EPA has found much of TSCA difficult to implement—hampering the agency's ability to obtain certain chemical data or place limits on chemicals. Of the thousands of chemicals listed for commercial use in the United States, EPA has used its authority to limit or ban five chemicals since TSCA was enacted. In 2009, EPA announced TSCA reform principles to inform ongoing efforts in Congress to strengthen the act. At that time, EPA also initiated a new approach for managing toxic chemicals with the goal of ensuring the safety of chemicals using its existing authorities.

* This is an edited, reformatted and augmented version of the Highlights of GAO-13-249, a report to congressional requesters, dated March 2013.

GAO was asked to evaluate EPA's efforts to strengthen its management of chemicals. This report determines the extent to which (1) EPA has made progress implementing its new approach and (2) EPA's new approach positions it to achieve its goal of ensuring the safety of chemicals. GAO examined agency documents and TSCA rulemaking and interviewed agency officials and stakeholders from industry and environmental organizations.

What GAO Recommends

GAO recommends, among other things, that EPA develop strategies that address challenges impeding its ability to ensure chemical safety and identify the resources needed to so. EPA neither agreed nor disagreed with GAO's recommendations.

What GAO Found

Since 2009, the Environmental Protection Agency (EPA) has made progress implementing its new approach to managing toxic chemicals under its existing Toxic Substances Control Act (TSCA) authority; particularly by increasing efforts to obtain chemical toxicity and exposure data and initiating chemical risk assessments—which EPA uses, along with other information, to decide what regulatory or other actions, if any, are warranted. The results of EPA's data collection activities, in most cases, have yet to be realized, and it may take several years before EPA obtains much of the data it is seeking. Also, EPA has not pursued some opportunities to obtain chemical data that companies submit to foreign governments or to obtain data from chemical processors that prepare chemical substances after their manufacture for distribution in commerce—some of which could help support the agency's risk assessment activities. Of the 83 chemicals EPA has prioritized for risk assessment, it initiated 7 assessments in 2012 and plans to start 18 additional assessments in 2013 and 2014. However, it may take several years to complete these initial risk assessments and, at the agency's current pace, over a decade to complete all 83, especially as EPA does not have the toxicity and exposure data needed for 58 of the 83 chemicals prioritized for risk assessment. In addition to its risk assessment activity, EPA has initiated other actions—such as increasing review of certain new uses of chemicals—that may discourage the use of these chemicals, but it is too early to tell whether these actions will reduce chemical risks.

It is unclear whether EPA's new approach to managing chemicals within its existing TSCA authorities will position the agency to achieve its goal of ensuring the safety of chemicals. EPA officials said that the agency's new approach, summarized in its 2012 *Existing Chemicals Program Strategy,* is intended to guide EPA's efforts to assess and control chemicals in the coming years. However, EPA's strategy, which largely focuses on describing activities EPA has already begun, does not include leading federal strategic planning practices that could help guide its effort. Specifically, EPA has not defined strategies that address challenges—many of which are rooted in TSCA's regulatory framework—that may impede EPA's ability to meet its long-term goal of ensuring chemical safety. Specifically, EPA has not clearly articulated how it will address challenges associated with obtaining toxicity and exposure data needed for risk assessments and placing limits on or banning chemicals under existing TSCA authorities. In addition, EPA's strategy does not describe the resources needed to execute its new approach. For example, EPA's strategy does not identify roles and responsibilities of key staff or offices or identify staffing levels or costs associated with conducting the activities under its new approach. Without a plan that incorporates leading strategic planning practices, EPA cannot be assured that its new approach to managing chemicals, as described in its *Existing Chemicals Program Strategy,* will provide a framework to effectively guide its effort. Consequently, EPA could be investing valuable resources, time, and effort without being certain that its efforts will bring the agency closer to achieving its goal of ensuring the safety of chemicals.

ABBREVIATIONS

ADP	Action Development Process
ATO	antimony trioxide
BPA	bisphenol A
CAS	Chemical Abstracts Service
DCM	dichloromethane
EPA	Environmental Protection Agency
HHCB	1,3,4,6,7,8-Hexahydro-4,6,6,7,8,8,-hexamethylcyclopenta[g]-2- benzopyran
HPV	High Production Volume

IUPAC	International Union of Pure and Applied Chemistry
IRIS	Integrated Risk Information System
NCEA	National Center for Environmental Assessment
NMP	n-methylpyrrolidone
OIRA	Office of Information and Regulatory Affairs
OMB	Office of Management and Budget
OPPT	Office of Pollution Prevention and Toxics
ORD	Office of Research and Development
OCSPP	Office of Chemical Safety and Pollution Prevention
PBDE	polybrominated diphenyl ether
PCB	polychlorinated biphenyls
PVC	plastics/rubber compounding
TCE	trichloroethylene
TSCA	Toxic Substances Control Act

March 22, 2013

The Honorable Barbara Boxer
Chairman
Committee on Environment and Public Works
United States Senate

The Honorable Tom Udall
Chairman
Subcommittee on Superfund, Toxics and Environmental Health
Committee on Environment and Public Works
United States Senate

The Honorable Frank Lautenberg
United States Senate

Tens of thousands of chemicals are listed with the Environmental Protection Agency (EPA) for commercial use in the United States, with an average of 600 new chemicals listed each year. EPA's ability to effectively implement its mission of protecting public health and the environment depends on credible and timely assessments of the risks posed by toxic chemicals. In 1976, Congress passed the Toxic Substances Control Act (TSCA) to provide EPA with the authority to obtain more information on chemicals and to regulate those chemicals that EPA determines pose unreasonable risks to human health or the environment. TSCA authorizes EPA to review chemicals already in commerce (existing chemicals) and chemicals yet to enter commerce (new chemicals). The scope of TSCA includes those chemicals manufactured, imported, processed,[1] distributed in commerce, used, or disposed of in the United States but excludes certain substances regulated under other laws.[2] TSCA also specifies when EPA may publicly disclose chemical information it obtains from chemical companies and provides that chemical companies can claim certain information, such as data disclosing chemical processes, as confidential business information.[3]

We have reported in the past, among other things, that EPA has found many of the provisions of TSCA difficult to implement—which has hampered the agency's ability to obtain certain chemical data, place limits on chemicals, and manage company assertions of confidentiality. Specifically, EPA has found it difficult to obtain adequate information on chemical toxicity and exposure[4] because TSCA does not require companies to provide this information and, instead, requires EPA to demonstrate that chemicals pose certain risks before it can ask for such information. Without adequate information on chemical toxicity and exposure, EPA is unable to assess the risks posed by many chemicals.

Even when EPA has had adequate information on toxicity and exposure, the agency has had difficulty demonstrating, under the standards required by TSCA, that harmful chemicals pose an unreasonable risk and, therefore, should be banned or have limits placed on their production or use. Consequently, EPA has used its authority to limit or ban the use of five chemicals since TSCA was enacted in 1976. The agency last used this authority in 1990. In addition, while EPA has reported that 95 percent of information it receives on new chemicals contains assertions of confidentiality, EPA officials have stated that they have not had the resources that would be needed to investigate and, as appropriate, challenge such claims.

In our past reports, we have suggested that Congress consider making statutory changes to strengthen EPA's authority to obtain toxicity information from the chemical industry and establish a framework for taking action that is less burdensome for EPA. In addition, we have identified a number of options that could strengthen EPA's ability to regulate harmful chemicals under TSCA.[5] For the past several years, congressional committees annually have considered legislation aimed at reforming TSCA, but Congress has not passed such legislation. We have also made recommendations to EPA. In June 2005, we reported that EPA had not used its TSCA authority to obtain information that U.S. companies submit to foreign governments, which may contain important information on chemical toxicity, and we recommended that EPA promulgate a rule requiring that companies submit copies to the agency of any health and safety studies, as well as other information concerning the environmental and health effects of chemicals that they submit to foreign governments.[6] We also recommended that the agency improve and validate its models for assessing and predicting the risks of chemicals and revise its regulations to require companies to reassert confidentiality claims within a certain period. EPA implemented our recommendation to improve its models. EPA did not disagree with our recommendations regarding obtaining health and safety studies and other information that companies submit to foreign governments and requiring companies to reassert confidentiality claims, but it provided substantive comments and has not fully implemented these recommendations. For this and other reasons, in 2009, we added EPA's processes for assessing and controlling toxic chemicals to our list of programs at high risk of waste, fraud, abuse, and mismanagement.[7]

Interest in reforming TSCA has heightened in recent years and, in 2009, EPA announced principles for reforming TSCA to help inform efforts under way in Congress. These principles include, among other things, goals for reforming TSCA so that (1) EPA would have clear authority to establish safety standards that are based on scientific risk assessments; (2) manufacturers would be required to provide sufficient toxicity, exposure, and use data for a chemical to support a determination by EPA that the chemical meets the safety standard; (3) EPA would have clear authority to take regulatory or other actions when chemicals do not meet the safety standard, with flexibility to take into account a range of considerations, including children's health, economic costs, social benefits, and equity concerns; (4) EPA would have authority to set priorities for conducting safety reviews on existing chemicals based on relevant risk and exposure considerations; and (5) EPA would receive a sustained source of funding from manufacturers of chemicals to support the costs of agency implementation, including the review of information provided by manufacturers.

Along with the announcement of these principles in 2009, EPA initiated a new approach to managing chemicals within the limits of existing authorities—which, according to agency documents, will transition the agency from an approach dominated by voluntary data submissions by industry to a more proactive approach in which EPA will use its data collection and other rulemaking authorities under TSCA to ensure chemical safety. In this context, you asked us to evaluate EPA's recent efforts to strengthen its management of chemicals.

Our objectives were to determine the extent to which (1) EPA has made progress implementing its new approach to managing toxic chemicals under its existing TSCA authority and (2) EPA's new approach positions the agency to achieve its goal of ensuring the safety of chemicals.

To address these objectives, we identified and reviewed EPA's efforts to obtain and analyze toxicity and exposure data, prioritize and perform risk assessments on chemicals, initiate regulatory and other actions to reduce risks; and make more chemical information—particularly data classified as confidential business information—available to the public. As part of this work, we reviewed EPA's TSCA rulemaking actions for the last 10 years and examined EPA's key policy, planning, and strategy documents. We interviewed officials from EPA's Office of Chemical Safety and Pollution Prevention, including its Office of Pollution Prevention and Toxics, regarding EPA's efforts to strengthen its management of chemicals. We interviewed officials from EPA's Office of Research and Development and Office of Policy relating to EPA's use of new analytical methods and tools and EPA's rulemaking process, respectively.

We also interviewed representatives from stakeholder groups, such as industry associations and environmental organizations. A more detailed description of our scope and methodology can be found in appendix I.

We conducted this performance audit from December 2011 to March 2013 in accordance with generally accepted government auditing standards. Those standards require that we plan and perform the audit to obtain sufficient, appropriate evidence to provide a reasonable basis for our findings and conclusions based on our audit objectives. We believe that the evidence obtained provides a reasonable basis for our findings and conclusions based on our audit objectives.

BACKGROUND

This section discusses (1) chemical risk assessments, (2) the structure of TSCA and prior findings that have impeded EPA's ability to assess and control toxic chemicals, and (3) EPA's new approach to managing chemicals within the limits of existing authorities.

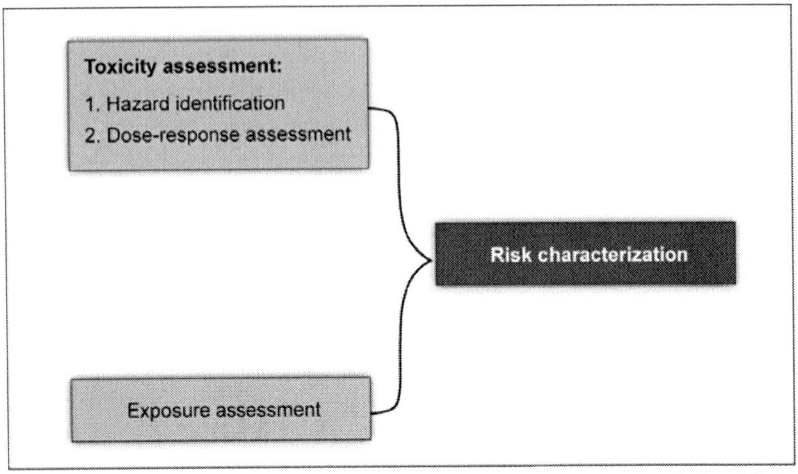

Sources: GAO presentation of the risk assessment component of the National Academies' risk assessment model used by EPA.

Figure 1. Human Health Chemical Risk Assessment Model Used by EPA.

Chemical Risk Assessments

EPA's ability to effectively implement its mission of protecting public health and the environment depends on credible and timely assessments of the risks posed by chemicals. EPA assesses the risks chemicals pose to human health (human health risk assessments) as well as to the environment (ecological risk assessments).

The human health risk assessment model EPA uses (see fig. 1) includes steps to (1) identify a chemical's toxicity, or hazardous properties, which are the potential noncancer and cancer human health effects of exposure to a chemical; (2) assess the dose-response relationship between exposure to a chemical and the resultant health effects, which describes the magnitude of hazard for potential noncancer effects and increased cancer risk; (3) assess the extent of human exposure to the chemical; and (4) characterize risk by determining the probability that populations or individuals so exposed to the chemical will be harmed and to what degree. Such assessments are the cornerstone of scientifically sound environmental decisions, policies, and regulations under a variety of statutes, including TSCA.

For some, but not all, chemicals, EPA conducts the first two steps of its chemical risk assessment model—that is, the hazard identification and dose-response assessment—under its Integrated Risk Information System (IRIS) program.[8] Taken together, these two steps are commonly referred to as toxicity assessments. EPA created the IRIS program in 1985 to help develop consensus opinions within the agency about the human health effects from chronic exposure to chemicals, and it is the only federal program that provides qualitative and quantitative assessments of both cancer risks and noncancer effects of chemicals. EPA's IRIS program—managed by EPA's National Center for Environmental Assessment (NCEA) within the Office of Research and Development (ORD)—develops new IRIS toxicity assessments and updates existing IRIS assessments if revisions are warranted on the basis of newly published peer-reviewed studies. For many chemicals, however, IRIS toxicity assessments are not available, current, or applicable to EPA's program offices—including the Office of Pollution Prevention and Toxics, which is responsible for implementing TSCA. Therefore, these offices, in some cases, prepare their own toxicity assessments.

EPA program offices, including the Office of Pollution Prevention and Toxics, combine information from IRIS toxicity assessments or other toxicity assessments with the results from chemical exposure assessments to develop human health risk assessments, which characterize the risks posed by chemicals, and make risk management decisions. Risk management, as opposed to risk assessment, involves integrating the risk assessment information with other information—such as economic information on the costs and benefits of mitigating a risk, technological information on the feasibility of managing the risk, and the concerns of various stakeholders—to determine whether the health risks identified in a chemical risk assessment warrant EPA taking regulatory or other risk management actions.

In the case of EPA's ecological risk assessment model, EPA's guidelines suggest a three-step process consisting of (1) problem formulation, (2) analysis, and (3) risk characterization, rather than the four-step process used for human health risk assessments. While for a human health risk assessment EPA is primarily concerned with a chemical's toxicity to humans, for an ecological risk assessment, the agency might consider a range of adverse effects on natural resources (e.g., crops, livestock, commercial fisheries, and forests), wildlife (including plants), aesthetic values (e.g., clear air in a national park), and recreational opportunities. According to EPA's ecological risk assessment guidelines, an ecological risk assessment is a process that evaluates the likelihood that adverse ecological effects may occur or are occurring as a result of exposure to one or more environmental stressors, such as chemicals, disease, invasive species, and climate change, and one stressor or many stressors may be considered. EPA's guidance on ecological risk assessments focuses on stressors and adverse ecological effects generated or influenced by human activity that could be addressed by the agency's risk management decisions.

TSCA's Structure and Prior Findings on Challenges Posed by TSCA

EPA's authority to ensure that chemicals in commerce do not present an unreasonable risk of injury to health or the environment is established in five major sections of TSCA. The purpose and application of these sections are shown in table 1 and described in further detail below.

Under the provisions for chemical testing in section 4 of TSCA, EPA can promulgate rules to require chemical companies to test potentially harmful chemicals for their health and environmental effects. However, EPA must first determine that testing is warranted based on some toxicity or exposure information. Specifically, to require such testing, EPA must find that a chemical (1) may present an unreasonable risk of injury to health or the environment or (2) is or will be produced in substantial quantities and that either (a) there is or may be significant or substantial human exposure to the chemical or (b) the chemical enters or may reasonably be anticipated to enter the environment in substantial quantities. EPA must also determine that there are insufficient data to reasonably determine or predict the effects of the chemical on health or the environment and that testing is necessary to develop such data. As we have previously reported, EPA has found this authority to be difficult, time-consuming, and costly to use.[9] The structure of section 4 of the act places the burden on EPA to demonstrate a need for data on a chemical's toxicity—rather than on a company—to demonstrate that a chemical is safe.

Under the provisions for new chemical review and significant new use rules in section 5 of TSCA, chemical companies are to notify EPA at least 90 days before beginning to manufacture a new chemical (premanufacture notice review). Section 5 also allows EPA to promulgate significant new use rules, which require companies to notify EPA at least 90 days before beginning to manufacture a chemical for certain new uses or in certain new ways (significant new use notice review). Such rules require existing chemicals to undergo the same type of review that new chemicals undergo. For example, EPA may issue a significant new use rule if it learns that a chemical that has previously been processed as a liquid is now being processed as a powder, which may change how workers are exposed to the chemical. For both new chemicals and significant new use reviews, the required notification to EPA must include certain information on chemical identity, use, production volume, and worker exposure, among other information. EPA has 90 days to review the information in the notice and identify potential risks. If EPA takes no action, manufacturing may commence.[10] Section 5 of the act also authorizes EPA to maintain a list of chemicals—called the chemicals of concern list—that present or may present an unreasonable risk of injury to health or the environment.

Under the provisions for chemical regulation in section 6 of TSCA, EPA is to apply regulatory requirements to chemicals for which EPA finds a reasonable basis exists to conclude that the chemical presents or will present an unreasonable risk of injury to health or the environment. To adequately protect against a chemical's risk, EPA can promulgate a rule that bans or restricts the chemical's production, processing, distribution in commerce, disposal, or use or requires warning labels be placed on the chemical. Under TSCA, EPA must choose the least burdensome requirement that will adequately protect against the risk. In promulgating a rule, EPA must consider and publish a statement in the *Federal Register* regarding the following: (1) the effects of the chemical on health and the environment and the magnitude of human and environmental exposure; (2) the benefits of the chemical for various uses and the availability of substitutes for those uses; and (3) the reasonably ascertainable consequences of the rule, after consideration of the effect on the national economy, small businesses, technological innovation, the environment, and public health.

Table 1. Purpose and Application of TSCA's Major Sections

Section	Purpose	Provides EPA with a mechanism to:
4	Chemical testing	Require companies to develop toxicity data under certain circumstances
5	New chemical review and significant new use rules	Review existing information, including exposure and toxicity data for new chemicals and certain new uses of existing chemicals
6	Chemical regulation	Limit or ban a chemical, among other controls
8	Industry reporting of chemical data	Obtain existing data, including exposure and toxicity data
14	Disclosure of chemical data	Disclose certain data provided to or obtained by EPA while also protecting confidential business information

Source: GAO analysis of TSCA.

As we have reported in the past, EPA has had difficulty demonstrating that chemicals should be banned or have limits placed on their production or use under section 6.[11] Since Congress enacted TSCA in 1976, EPA has issued regulations under section 6 of the act to ban or limit the production or restrict the use of five existing chemicals or chemical classes out of tens of thousands of chemicals listed for commercial use on the agency's TSCA inventory—polychlorinated biphenyls (PCB), fully halogenated chlorofluoroalkanes, dioxin, asbestos, and hexavalent chromium.[12] EPA's 1989 asbestos rule illustrates the difficulties EPA has had in issuing regulations to control existing chemicals. In 1979, EPA started considering rulemaking on asbestos. After concluding that asbestos was a potential carcinogen at all levels of exposure,[13] EPA promulgated a rule in 1989 prohibiting the future manufacture, importation, processing, and distribution of asbestos in almost all products. Some manufacturers of asbestos products filed suit against EPA, arguing, in part, that the rule was not promulgated on the basis of substantial evidence regarding unreasonable risk.

In 1991, the Fifth Circuit Court of Appeals ruled for the manufacturers and returned parts of the rule to EPA for reconsideration. In reaching this conclusion, the court found that EPA did not consider all necessary evidence and failed to show that the control action it chose was the least burdensome reasonable regulation. The court further criticized EPA for banning a product for which no substitutes were currently available. EPA has not proposed a new section 6 rule since the court's ruling in 1991.

Under the provisions for industry reporting of chemical data in section 8(a), EPA is to promulgate rules under which chemical companies must maintain records and submit such information as the EPA Administrator reasonably requires. This information can include, among other things, chemical identity, categories of use, production levels, by-products, existing data on adverse human health and environmental effects, and the number of workers exposed to the chemical, to the extent such information is known or reasonably ascertainable.

Under section 8(a), EPA issues rules to collect additional or updated information on certain TSCA inventory chemicals. For example, in August 2011, EPA finalized its TSCA Chemical Data Reporting rule (previously referred to as the Inventory Update Reporting Modifications Rule); the rule requires companies to report, among other things, exposure-related information, such as production volume and use data, on chemicals manufactured or imported over a certain volume per year. EPA uses production volume and use data reported by companies as a proxy to estimate the extent of human exposure to the chemicals. In addition, section 8(d) provides EPA with the authority to promulgate rules under which chemical companies are required to submit lists or copies of existing health and safety studies to EPA. Section 8(e) requires chemical companies to report any information to EPA that reasonably supports a conclusion that a chemical presents a substantial risk of injury to health or the environment, unless the company has actual knowledge that EPA already has the information.

The provisions for disclosure of chemical data in section 14 of the act specify when EPA may disclose chemical information it obtains under TSCA. Chemical companies can claim certain information, such as data disclosing chemical processes, as confidential business information. EPA generally must protect confidential business information against public disclosure unless necessary to protect against an unreasonable risk of injury to health or the environment. Other federal agencies and federal contractors can obtain access to this confidential business information in order to carry out their responsibilities. EPA may also disclose certain data from health and safety studies.

Over the years, companies have classified much of the information they submitted to EPA as confidential business information, and, prior to 2009, EPA did not routinely challenge their assertions, citing resource constraints. As a result, the extent to which companies' confidentiality claims were warranted was unknown.

EPA's New Approach to Managing Chemicals

In September 2009, EPA initiated a new approach to managing chemicals that focuses largely on existing chemicals—that is, the approximately 84,000 chemicals already on the TSCA inventory.[14] EPA's new approach has evolved over time and includes a variety of different activities and initiatives. In February 2012, EPA summarized many of the activities it had initiated under its new approach in the agency's *Existing Chemicals Program Strategy*. Collectively, these activities address four areas: (1) collecting toxicity and exposure data, (2) conducting risk assessments, (3) discouraging the use of some chemicals, and (4) expanding public access to some chemical data.

EPA HAS MADE PROGRESS IMPLEMENTING ITS NEW APPROACH TO MANAGING CHEMICALS, BUT, IN MOST CASES, RESULTS HAVE YET TO BE REALIZED

EPA has made progress implementing its new approach to managing toxic chemicals under its existing TSCA authority—particularly by increasing efforts to (1) obtain toxicity and exposure data, (2) assess risks posed by chemicals, (3) discourage the use of some chemicals, and (4) expand public access to some chemical information. However, the results of EPA's activities, in most cases, have yet to be realized.

EPA Has Increased Efforts to Collect Data on Toxicity and Exposure, but It May Take Several Years to Produce Results

Since 2009, EPA has increased its efforts to collect toxicity and exposure data. Specifically, since 2009, EPA has required companies to test 34 chemicals and provide EPA with the resulting toxicity and other data. In addition, in 2011, EPA announced, but has yet to finalize, plans to require testing for 23 additional chemicals.[15] By comparison, EPA promulgated test rules for 197 chemicals from the time TSCA was enacted in 1976 until 2009 when the agency undertook its new approach to managing chemicals.[16] The 57 chemicals that are part of EPA's current and proposed testing requirements were identified but not sponsored as part of the agency's1998 voluntary effort to obtain testing data from companies on chemicals produced or imported at high volumes (i.e., amounts of 1 million pounds or more a year).[17]

Due to requirements under TSCA that place the burden of developing toxicity data on EPA, rather than on industry, and because EPA's past efforts to obtain these data voluntarily were not successful, EPA proposed or promulgated rules to require chemical companies to test these 57 chemicals. However, because rules can take years to finalize and additional time for companies to execute, EPA has yet to obtain much of the information it has been seeking. According to EPA officials, it can take, on average, 3 to 5 years for the agency to promulgate a test rule and an additional 2 to 2 ½ years for the companies to provide the data once EPA has requested them.

In addition, toxicity data eventually obtained on the 57 chemicals may not, in all cases, be sufficient for EPA to conduct a risk assessment (i.e., characterize risk by determining the probability that populations or individuals so exposed to a chemical will be harmed and to what degree). EPA officials told us that much of the chemical toxicity information obtained previously through its 1998 voluntary effort to obtain testing data from companies is considered "screening level" information. That is, the information was collected to identify a chemical's potential hazards to human health and the environment, but it was not intended to be the basis for assessing whether a chemical poses an unreasonable risk of injury to human health or the environment, according to agency documents describing the program. EPA's efforts since 2009 to require companies to test chemicals is based on testing parameters similar to those used under its voluntary effort and thus may produce similar basic screening level data.[18]

With regard to exposure data, in August 2011, EPA revised its periodic chemical data reporting requirements to obtain exposure-related information for a greater number of chemicals. Under the revised requirements, EPA (1) lowered the reporting thresholds, in some cases,[19] which will allow it to look at exposure scenarios for a larger number of chemicals than in the past and (2) shortened the reporting cycle from every 5 years to every 4 years. In addition, starting in 2016, the revised requirements for reporting will be triggered when companies exceed applicable production thresholds in any year during the 4-year reporting cycle. Previously, the reporting requirement was triggered only if production levels were exceeded during the reporting year. According to EPA officials, this change was important because, under the previous requirement, production volumes of chemical substances fluctuated above and below reporting thresholds in different reporting periods, resulting in a change of approximately 30 percent in the composition of the chemical substances reported as being produced from one reporting period to the next. EPA received the first batch of exposure-related data under the new reporting requirements for approximately 8,000 chemicals in August 2012, and EPA officials told us they expect to begin analyzing these data in the coming months. After receiving and analyzing these exposure-related data, according to agency documents, EPA plans to determine how to use the information to identify additional data collection needs or identify chemicals that may warrant further review or risk assessment.

Even with the steps it has taken since 2009 to increase the toxicity and exposure data it collects, EPA has not pursued all opportunities to obtain chemical data. In particular, EPA has not broadly sought toxicity and exposure data that companies submit to the European Chemicals Agency on chemicals that the companies manufacture or process in, or import to, the United States.[20] Under the European Union's chemicals legislation, the European Chemicals

Agency may share information it receives from chemical companies with foreign governments in accordance with a formal agreement concluded between the European Community and the foreign government, but EPA has not pursued such an agreement. According to EPA, it has an informal agreement, or Statement of Intent, for cooperation and sharing of information with European Chemicals Agency, and had hoped that such an agreement would allow for the sharing of detailed studies, beyond the summaries made publically available by European Chemicals Agency. In addition, EPA has not issued a rule under section 8 of TSCA requiring companies to provide EPA with the information provided to the European Chemicals Agency. EPA officials told us that the agency has not sought to obtain chemical data—from either the European Chemicals Agency or companies directly—because it does not believe that this would be the best use of EPA or industry resources. They also said that it is unclear whether these data would be useful to EPA. EPA officials believe it is a more effective use of resources to gain access to data, as needed, on a case-by-case basis from chemical companies.

EPA has also not used its authority to obtain exposure-related data from chemical processors that prepare chemical substances or mixtures, after their manufacture, for distribution in commerce.[21] Specifically, EPA has not issued a rule under section 8 of TSCA to extend its periodic chemical data reporting requirements to chemical processors. EPA's chemical data reporting requirements apply only to chemical manufacturers and importers. However, chemical processors are often the downstream users of chemical substances produced by chemical manufacturers and, therefore, may be in a better position to understand end users' exposure scenarios. EPA's principles for TSCA reform stress the importance of these data, stating that, "EPA's authority to require submission of use and exposure information should extend to downstream processors…" In addition, EPA officials told us that data from processors would provide the agency with a better understanding of potential exposure to chemicals, for example, from consumer products such as those designed for children, and that these data are necessary to conduct chemical risk assessments and make risk management decisions on potentially harmful chemicals. Nonetheless, the same EPA officials told us that the agency has not sought to collect such data from all processors because the agency does not currently have the resources to receive, store, and analyze the additional data. Instead, according to EPA officials, the agency has worked on a case-by-case basis with processors and processor associations to ask them to voluntarily submit data.

With regard to obtaining toxicity and exposure data submitted to the European Chemicals Agency and exposure-related data from chemical processors, EPA officials told us that they have considered using EPA's subpoena authority under TSCA section 11(c) to obtain the information if they are unable to obtain it voluntarily—which is an approach EPA has not frequently used before.[22] EPA officials said that they recognized that rules under section 8 of TSCA could be fashioned in such a way as to establish general access to information while also providing EPA with the flexibility to request the information as needed. However, according to EPA officials, TSCA section 11(c) authority may be appropriate in situations where EPA is seeking specific data and information from a specific entity, rather than a more general unknown population where rulemaking might be more appropriate. EPA has yet to take such action under its new approach to managing chemicals, however, and it is unclear whether such an approach would provide the agency with timely access to needed information. Without access to the data that companies have submitted to the European Chemicals Agency and by not pursuing exposure-related data from processors, regardless of the mechanism used, EPA is missing an opportunity to collect data that it has identified as an essential part of assessing chemical risk and future chemical regulation.

EPA Has Begun Assessing Chemical Risks, but It Is Too Early to Tell What, If Any, Risk Management Actions Will be Taken

In February 2012, EPA announced a plan that identified 83 existing chemicals for risk assessment—known as the TSCA Work Plan.[23] From this list of 83 chemicals, EPA's Office of Pollution Prevention and Toxics—the office responsible for implementing TSCA—initiated risk assessments for 7 chemicals in 2012[24] and announced plans to start risk assessments during 2013 and 2014 for 18 additional chemicals. While EPA's effort to initiate TSCA-related risk assessments represents a significant increase in risk assessment activity,[25] it may be years before EPA initiates risk management actions to reduce any chemical risks identified in these assessments. In January 2013, EPA released 5 draft risk assessments for public comment; from the 7 chemicals for which it initiated risk assessments in 2012.[26] The two remaining assessments are to be released for public comment at a later date, according to EPA's website. EPA officials told us that, upon completion of the public comment periods and its external peer-review process, all 7 risk assessments will be finalized early in 2014. After EPA completes each risk assessment, according to EPA documents, the agency plans to determine what risk management actions, if any, are warranted to address identified risks. As discussed previously, before EPA can determine whether regulatory or other risk management action is warranted, the agency would need to consider other factors—such as economic information on the costs and benefits of mitigating the risk, technological information on the feasibility of managing the risk, and the concerns of various stakeholders such as industry and environmental organizations—which could require additional time and resources beyond completing risk assessments. Moreover, assuming EPA meets its target for completing these assessments and initiating new assessments, at its current pace, it would take EPA at least 10 years to complete risk assessments for the 83 chemicals in the TSCA Work Plan.

In addition, it is not clear that EPA can maintain its current pace given that it currently does not have the toxicity and exposure data it will need to conduct risk assessments for all of the 83 chemicals in its TSCA Work Plan, and it is unclear how or when EPA will obtain these data. According to EPA officials and agency documents, of the 83 chemicals identified in its TSCA Work Plan, the agency has started or plans to start risk assessments on the 25 chemicals for which it has well-characterized toxicity and exposure data. Before EPA can initiate risk assessments for the remaining 58 chemicals, the agency will need to identify the toxicity and exposure data it needs and then obtain them. According to agency officials, to obtain the toxicity data needed, EPA may need to promulgate rules to require companies to perform additional testing on some of these chemicals. However, EPA has not clearly articulated how or when it plans to obtain these needed data. Moreover, without exposure-related data, such as those potentially available from chemical processors, EPA may still be missing the data necessary to conduct risk assessments.

The type and scope of each of EPA's planned risk assessments are also unclear. For the 76 risk assessments in the TSCA Work Plan that EPA has yet to initiate, it is not clear what type of risk assessment EPA will perform. That is, whether a risk assessment will focus on human health, ecological hazards, or both—or how broadly or narrowly focused a risk assessment will be in terms of assessing exposure scenarios. The type and scope of a risk assessment will, in part, affect the risk management options available to EPA—which can include a number of alternatives such as requiring special labeling and banning or limiting the use of a chemical. More is known about the type and scope of the 7 risk assessments EPA has initiated—particularly the 5 draft assessments released in January 2013. Examples are as follows:

- For two chemicals, EPA produced draft risk assessments that were focused on potential ecological hazards from specific chemical uses.[27] The draft risk assessments indicate a "low concern for ecological health."
- For two chemicals, EPA produced draft risk assessments that were focused on the human health hazards to consumers, including bystanders and workers, when exposed to these chemicals during paint stripping.[28] The draft risk assessments identified "potential concern for human health under specific exposure scenarios for particular uses."
- For one chemical, EPA produced a draft risk assessment focused on human health hazards to consumers, including bystanders and workers, from inhalation exposures when the chemical is used as a degreaser or a "clear protective coating spray" in the arts and crafts field.[29] The draft risk assessment identified "potential concern for human health under specific exposure scenarios for particular uses."
- For the two chemicals for which EPA has yet to release draft risk assessments, EPA plans to focus primarily on releases to the environment from the processing and use from metalworking fluids and plastics/rubber (PVC) compounding.[30]

It is also not clear where EPA will obtain and who will develop the toxicity assessments needed to support its planned risk assessments—for example, whether IRIS toxicity assessments will be used, in whole or in part, or if some other type of assessment will be developed or used. According to EPA officials with the Office of Pollution Prevention and Toxics, they plan to incorporate information from IRIS toxicity assessments to the extent such information is available, recent, and relevant. For example, two of the three draft risk assessments discussed above that focused on human health hazards incorporated information from IRIS toxicity assessments, according to agency officials.[31] However, as we have reported previously,[32] IRIS toxicity assessments are not available or current for many chemicals, including many of the remaining chemicals in the TSCA Work Plan. Officials with the Office of Pollution Prevention and Toxics have not requested IRIS assessments through the formal IRIS nomination process for any of the 83 chemicals prioritized for risk assessment in the TSCA Work Plan but, according to IRIS program officials, they are working to find other options for assessing toxicity when IRIS assessments are not available, recent, or applicable. According to officials with the Office of Pollution Prevention and Toxics, they have not requested IRIS assessments for these 83 chemicals because IRIS toxicity assessments often take years to complete, among other reasons. In addition, these officials noted that IRIS assessments are generally used to estimate risks associated with continuous exposures to a pollutant in the air or water rather than the intermittent exposures that workers and consumers are subject to from chemicals contained in products.[33] While IRIS toxicity assessments are not essential for conducting a risk assessment, EPA officials have described IRIS assessments as the premier national and international source for qualitative and quantitative chemical risk information. In addition, no other federal toxicity assessment program has internal and external peer-review processes that are as rigorous according to EPA's recent *Human Health Risk Assessment Strategic Research Action Plan*.[34] The rigor associated with an IRIS assessment may be an important consideration in defending regulations that ban or limit the use of a chemical.

EPA Has Taken Actions That May Discourage the Use of Certain Chemicals, but It Is Too Early to Tell Whether These Actions Will Reduce Chemical Risk

Given the difficulty that EPA has faced in the past using section 6 of TSCA to ban existing toxic chemicals or place limits on their production or use, EPA officials told us that the agency generally considers section 6 authority only after exhausting all other available options. As such, since 2009, EPA has taken other actions that may discourage the use of certain chemicals by (1) making greater use of significant new use rules under section 5 and (2) proposing actions that use its TSCA authority in new ways. However, it is too early to tell whether some of these actions will reduce chemical risks.

Our analysis of TSCA rulemaking from 2009 to 2012 shows that EPA has quadrupled its issuance of significant new use rules since 2009. From 2009 to 2012, EPA issued significant new use rules affecting about 540 chemicals, about 25 percent of all 2,180 chemicals subject to significant new use rules issued by EPA since 1976. According to EPA officials, for chemicals subject to significant new use rules, EPA typically recommends that companies submit testing information when they notify EPA of their intent to manufacture or process chemicals subject to such rules, which enables EPA to better evaluate the potential risks associated with the new use. According to EPA officials, this approach allows the agency to "chip away" at chemicals that may pose risks to human health and the environment. Such recommendations may discourage companies from pursuing new uses of existing chemicals that may pose health or environmental risks either because testing itself can be expensive, or because the testing recommendation suggests that the agency may consider banning or limiting the manufacture or production of the chemical on the basis of that testing. One industry stakeholder told us that while EPA has not directly regulated chemicals through significant new use rules, EPA's use of these rules has deterred companies from pursuing new uses of these chemicals.

EPA has also proposed actions that use its TSCA authority in new ways and that, according to agency officials, are intended to discourage the use of certain chemicals that may pose health or environmental risks. However, it is too early to assess the impact of EPA's proposed actions because they have yet to be finalized. In addition, in some cases, the Office of Management and Budget (OMB) has not met the established 90- day time for reviewing EPA's proposed actions—which has increased the time frames for finalizing them.[35] These proposed actions include the following:

- *Creating "Chemicals of concern" list.* In May 2010, EPA announced that it intended to create a list of chemicals that present or may present "an unreasonable risk of injury to health or the environment." EPA has had the authority to create such a list under section 5 of TSCA since its enactment in 1976 but has never attempted to use this authority. EPA submitted the list, which consists of three groups of chemicals, for review by OMB in May 2010.[36] Although the period for OMB review is generally limited by executive order to 90 days,[37] as of December 2012, EPA's proposed "chemicals of concern" list has been under review at OMB for over 900 days. Stakeholders we interviewed had differing perspectives on EPA's proposed use of this list. One said that EPA did not assert clear criteria for a chemical's inclusion on the list and that being on the list has the effect of blacklisting a chemical and negatively impacting the market. Alternatively, another stakeholder we spoke with said that the list is analogous to the European Union's candidate list, which acts to provide companies sufficient time to respond to possible future regulation. The list thus gets the market moving to either defend the safety of the chemical or to get the chemical out of production.
- *Pairing of test and significant new use rules.* In December 2010, EPA submitted to OMB for review a proposal to pair testing rules with significant new use rules for the first time. Specifically, EPA has

proposed single rules that combine provisions requiring companies to develop toxicity and other data with provisions requiring companies to provide data for new uses of chemicals. EPA has proposed using this approach in two cases. In one case, for example, EPA proposed this approach for certain Polybrominated Diphenyl Ethers (PBDE)—flame retardants that are being voluntarily phased out, effective December 2013. Under the proposed rule, any new use of the chemical after it has been phased out would qualify as a significant new use, triggering a testing requirement. According to EPA officials, the pairing of these types of rules is intended to discourage new uses of certain chemicals that may pose a risk to human health or the environment and create a disincentive for companies to continue current use of the chemical— something EPA has not done before. OMB's review of this proposal took 422 days and was completed on February 15, 2012. As of January 2013, this rule has yet to be finalized.

- *Extending significant new use rules to articles.* Since 2009, EPA has made increasing use of its ability to subject chemicals contained in certain products, or "articles," such as furniture, textiles, and electronics, to significant new use rules. Generally, those who import or process a substance as part of a product are exempted from compliance with a significant new use rule. EPA's proposals would eliminate this exemption for certain chemicals.[38] Some stakeholders stated that the move to eliminate the article exemption for certain chemicals represents an attempt by EPA to regulate consumer products, not just the chemical substances they contain. Other stakeholders noted that EPA's ability to regulate potentially harmful chemicals is diminishing given their increased production outside of the United States, and that an increasing focus on articles allows EPA to minimize exposure to potentially harmful chemicals by targeting chemicals in imports as part of consumer products.[39]

EPA Has Expanded Public Access to Some Chemical Information

EPA has made progress in expanding public access to some chemical information—which according to EPA documents is an important underpinning of a credible chemical management program. When information is claimed as confidential business information, it limits EPA's ability to share it with state environmental agencies and foreign governments, which potentially limits the effectiveness of these organizations' environmental risk programs. Since 2009, EPA has made 617 formerly confidential chemical identities public.[40] By reviewing past claims of confidentiality and comparing them with more current reporting—such as reporting of periodic production and use data—EPA was able to identify chemicals for which companies were no longer making confidentiality claims. Since 2009, EPA has also made 783 previously unavailable health and safety filings available to the public after reviewing approximately 15,500 such filings.[41] In addition, EPA issued new policies regarding how it handles confidentiality claims and, according to EPA officials, has begun reviewing and challenging new confidentiality claims—including claims associated with chemical identity information listed on the TSCA inventory and in health and safety studies.[42] Further, as part of the 2011 Chemical Data Reporting rule, EPA included new upfront substantiation requirements for CBI claims related to processing and use information. According to EPA, the change has resulted in significantly decreased CBI claims for those data.

It Is Unclear Whether EPA's New Approach Will Position the Agency to Achieve Its Goal of Ensuring the Safety of Chemicals

It is unclear whether EPA's new approach to managing chemicals within its existing TSCA authorities will position the agency to achieve its goal of ensuring the safety of chemicals. EPA officials have said that the agency's new approach, initiated in 2009 and summarized in its 2012 *Existing Chemicals Program Strategy,* is intended to guide EPA's efforts to assess and control chemicals in the coming years. However, EPA's strategy, which largely focuses on describing activities EPA has already begun, does not discuss how it will address challenges that might impede its goal of ensuring chemical safety. Specifically, and as detailed in the list that follows, EPA's strategy does not discuss challenges associated with (1) obtaining toxicity and exposure data; (2) identifying the resources needed to execute EPA's new approach; and (3) banning or limiting the use of chemicals, given the agency's past difficulties with taking such actions.

- *Obtaining toxicity and exposure data.* EPA's strategy does not discuss how the agency will meet the challenge of obtaining the toxicity and exposure data it will need for conducting risk assessments for all 83 chemicals in its TSCA Work Plan. As discussed previously, EPA has not broadly sought toxicity and exposure data that companies submit to the European Chemicals Agency or exposure-related data from chemical processors and instead plans to obtain these data, as needed, on a case-by-case basis from chemical companies. However, the agency's strategy does not discuss how EPA would execute these plans or how the data obtained would be used to inform its ongoing or future risk assessment activities, if at all.
- *Identifying the resources needed.* EPA's strategy does not include a description of the resources needed to meet its goal of ensuring chemical safety. For example, EPA's strategy does not include a description of the resources needed to carry out risk assessment activities, even though risk assessment is a central part of EPA's effort to manage chemicals under its new approach. Specifically, EPA does not identify roles and responsibilities of key staff or offices—for example which office within EPA will develop the toxicity assessments needed to support its planned risk assessments—or identify staffing levels or cost associated with conducting its risk assessment activities. In response to our questions regarding resources, EPA officials provided us with an estimate of the staffing levels and contract support costs needed to conduct its first 7 risk assessments; however, EPA does not include this or any other information regarding staffing levels or cost in its strategy.[43] As discussed previously, at its current pace—which it may not be able to sustain—it will take more than a decade for EPA to complete the 83 risk assessments it has identified in its TSCA Work Plan. However, EPA officials were unable to tell us—and EPA strategy does not discuss—whether the agency has the capacity to accelerate the pace of its risk assessment activities. Further, the plan does not discuss how EPA will keep pace with the introduction of approximately 600 new chemicals each year, some of which may require risk assessments to determine whether they pose risks to health or the environment. Without a clear understanding of the resources needed to complete risk assessments and other activities identified in its strategy, EPA cannot be certain that its current funding and staffing levels are

sufficient to execute its new approach to managing chemicals under existing TSCA authorities.

- *Banning or limiting the use of chemicals.* EPA's strategy also does not discuss how it will address specific regulatory challenges that might impede the agency's ability to meet its goal of ensuring chemical safety. According to EPA officials, to demonstrate that a chemical presents or will present an unreasonable risk of injury to health or the environment—which would be required before EPA could ban or limit the chemical under section 6 of TSCA—EPA must develop or obtain toxicity and exposure data and then conduct a risk assessment. However, EPA officials told us that, even if EPA has substantial toxicity and exposure data and wants to protect the public against known risks, the agency is challenged in meeting the statutory requirement under TSCA to limit or ban chemicals. While agency planning documents related to EPA's new approach state that it would consider using its authority under section 6 of TSCA to limit or ban the use of some chemicals, EPA has yet to publicly take steps toward that end and has not articulated, in its strategy or elsewhere, how it would overcome the regulatory challenges it experienced in the past. As discussed previously, EPA officials told us that they would consider using the agency's authority under section 6 only after exhausting all other available options, but EPA's strategy does not discuss what other options the agency plans to pursue—for example, whether it plans to continue to rely on significant new use rules under section 5 of TSCA to discourage the use of certain chemicals. As a result, EPA cannot be assured that its new approach best positions the agency to ensure the safety of chemicals.

We have previously reported[44] that, when developing new initiatives, agencies can benefit from following leading practices for federal strategic planning.[45] Of these leading practices, it is particularly important for agencies to define strategies that address management challenges that threaten their ability to meet long-term goals.[46] Without a plan that incorporates leading strategic planning practices—particularly a plan that clearly articulates how EPA will address management challenges—EPA cannot be assured that it its new approach to managing chemicals, as described in its *Existing Chemicals Program Strategy,* will provide a framework to effectively guide its efforts. Consequently, EPA could be investing valuable resources, time, and effort without being certain that its efforts will bring the agency closer to achieving its goal of ensuring the safety of chemicals.

CONCLUSION

Since 2009, EPA has made progress implementing its new approach to managing toxic chemicals under its existing TSCA authority—particularly by increasing efforts to obtain toxicity and exposure data. However, due to requirements under TSCA that place the burden of developing toxicity data on EPA, rather than industry, and because promulgating the rules needed to obtain toxicity data from companies can take years to finalize, and additional time for companies to execute, EPA has yet to obtain much of the toxicity data it has been seeking. Also, EPA is in the process of analyzing exposure-related data it received in August 2012 and, therefore, is not yet in a position to use them to identify additional data collection needs or identify chemicals that may warrant further review or risk assessment. Moreover, while EPA's toxicity and exposure data collection efforts may be useful for identifying chemicals that are potentially harmful, it is unclear whether these data will be sufficient for conducting a risk assessment. According to EPA officials, EPA needs both toxicity and exposure data to conduct a risk assessment in order to demonstrate that a chemical presents or will present an unreasonable risk of injury to health or the environment—which would be required before EPA could ban or limit the use of a chemical under TSCA.

Even with the steps it has taken since 2009 to increase the toxicity and exposure data it collects, EPA has not pursued all opportunities to obtain chemical data. In particular, EPA has not sought (1) toxicity and exposure data that companies submit to the European Chemicals Agency or (2) generally pursued exposure-related data from chemical processors. That is, EPA has not pursued a formal agreement with the European Community or used its authority to promulgate rules under TSCA section 8 to require chemical companies to report chemical toxicity and exposure- related data they have submitted to the European Chemicals Agency. EPA has also not promulgated rules under TSCA section 8 to require chemical companies to report exposure-related data from processors to EPA. EPA officials said that they recognized that rules under section 8 of TSCA could be fashioned in such a way as to establish general access to information while also providing EPA with the flexibility to request the information as needed—but were considering a different approach. Agency officials told us that they have considered using

EPA's subpoena authority under TSCA section 11(c) to obtain the information—which is an approach EPA has not frequently used before. Regardless of the mechanism used, without access to the data that companies have submitted to the European Chemicals Agency and by not pursuing exposure-related data from processors, EPA is missing an opportunity to collect data that it has identified as an essential part of assessing chemical risk and future chemical regulation.

It is unclear whether EPA's new approach to managing chemicals within its existing TSCA authorities will position the agency to achieve its goal of ensuring the safety of chemicals because EPA has not clearly articulated how its strategy will address challenges that threaten an agency's ability to meet its goal—particularly, challenges associated with obtaining toxicity and exposure data needed for risk assessments and with EPA's ability to ban or limit the use of chemicals, given the agency's past difficulties with taking such actions. EPA officials have said that the agency's approach, summarized in its 2012 *Existing Chemicals Program Strategy,* is intended to guide EPA's efforts to assess and control chemicals in the coming years. However, EPA's strategy does not include leading federal strategic planning practices, such as defining strategies for addressing management challenges that might help the agency achieve its goal of ensuring the safety of chemicals. For example, EPA's strategy does not address challenges associated with (1) obtaining toxicity and exposure data for the 58 TSCA Work Plan chemicals for which it currently needs such data, (2) gaining access to toxicity and exposure data provided to the European Chemicals Agency, (3) working with processors and processor associations to obtain exposure-related data, (4) addressing specific regulatory challenges with banning or limiting the use of chemicals under section 6 of TSCA, and (5) identifying the resources needed to achieve its goal.

It is worth noting that many of the challenges that EPA faces are rooted in TSCA's regulatory framework. In our past reports, we have suggested that Congress consider making statutory changes to strengthen EPA's authority to obtain toxicity information from the chemical industry and establish a framework for taking action that is less burdensome for EPA; in addition, we have identified a number of options that could strengthen EPA's ability to regulate harmful chemicals under TSCA. Until Congress passes any such legislation, however, EPA can do more to improve its current effort to help bring the agency closer to achieving its goal of ensuring the safety of chemicals.

RECOMMENDATIONS FOR EXECUTIVE ACTION

To better position EPA to collect chemical toxicity and exposure-related data and ensure chemical safety under existing TSCA authority, while balancing its workload, we are recommending that the Administrator of EPA take the following three actions:

- Consider promulgating a rule under TSCA section 8, or take action under another section, as appropriate, to require chemical companies to report chemical toxicity and exposure-related data they have submitted to the European Chemicals Agency.
- Consider promulgating a rule under TSCA section 8, or take action under another section, as appropriate, to require chemical companies to report exposure-related data from processors to EPA.
- To better position EPA to ensure chemical safety under existing TSCA authority, direct the appropriate offices to develop strategies for addressing challenges that impede the agency's ability to meet its goal of ensuring chemical safety. At a minimum, the strategies should address challenges associated with:
    - obtaining toxicity and exposure data needed to conduct ongoing and future TSCA Work Plan risk assessments,
    - gaining access to toxicity and exposure data provided to the European Chemicals Agency,
    - working with processors and processor associations to obtain exposure-related data,
    - banning or limiting the use of chemicals under section 6 of TSCA and planned actions for overcoming these challenges—including a description of other actions the agency plans to pursue in lieu of banning or limiting the use of chemicals, and
    - identifying the resources needed to conduct risk assessments and implement risk management decisions in order to meet its goal of ensuring chemical safety.

AGENCY COMMENTS AND OUR EVALUATION

We provided a draft of this report to the Environmental Protection Agency for their review and comment. We received written comments from the Acting

Assistant Administrator of EPA's Office of Chemical Safety and Pollution Prevention. EPA also provided technical comments, which we incorporated into the report as appropriate.

In its written comments, EPA neither agreed nor disagreed with our findings and recommendations and instead stated that it appreciated the intent of our recommendations and will consider them as it further develops and implements the TSCA program. However, based on the comments the agency provided, it is unclear whether EPA intends to take any action toward implementing our recommendations. Specifically, it is unclear whether EPA plans to address our recommendations that the agency consider promulgating rules under TSCA section 8, or take action under another section, as appropriate, to require chemical companies (1) to report chemical toxicity and exposure-related data they have submitted to the European Chemicals Agency and (2) to report exposure- related data from processors to EPA. In its written comments, EPA states that it intends to pursue data submitted to the European Chemicals Agency from U.S. companies using voluntary or regulatory means as necessary but does not provide information on its planned approach to pursue such data. Consequently, the extent to which EPA plans to continue to rely on voluntary efforts to obtain the needed data is unclear. For example, it is not clear whether EPA intends to first ask companies to voluntarily comply with their data request and then, if that does not yield data, pursue regulatory action. In addition, it is unclear whether continuing to rely on voluntary efforts will provide the agency with timely access to needed data. As we noted in our report, EPA officials have recognized that rules under section 8 of TSCA could be fashioned in such a way as to establish general access to information while providing EPA with the flexibility to request the information as needed.

In addition, with regard to obtaining data from chemical processors, in its written comments, EPA states that downstream chemical processors have little exposure-relevant data—which suggests that it does not intend to implement that recommendation. This position, however, conflicts with previous statements by EPA officials and EPA's principles for TSCA reform, which state that, "EPA's authority to require submission of use and exposure information should extend to downstream processors..." In addition, EPA officials told us that data from downstream processors would provide the agency with a better understanding of potential exposure to chemicals, for example, chemical exposure from consumer products such as those designed for children. EPA officials also told us that these data are necessary to conduct chemical risk assessments and make risk management decisions on potentially harmful chemicals. In its written comments, EPA also states that the use of section 8 requires a lengthy and resource intensive rulemaking process. However, as previously noted, EPA officials have recognized that a rule under section 8 of TSCA could be promulgated to require chemical companies to report exposure-related data while also providing EPA with the flexibility to request the data as needed. In this way, should EPA identify a need for information from downstream processors, EPA would not have to go through the lengthy rulemaking process multiple times.

Regarding our recommendation to develop strategies for addressing challenges associated with obtaining toxicity and exposure data needed for risk assessments, and with EPA's ability to meet its goal of ensuring chemical safety, it is also unclear what action, if any, EPA intends to pursue. In its written comments, the agency states that it will not be able to meet the goal of ensuring chemical safety now and into the future without legislative reform and, until then, EPA plans to utilize its *Existing Chemicals Program Strategy*. However, as discussed in this report, EPA's *Existing Chemicals Program Strategy*, which largely focuses on describing activities the agency is already undertaking and is therefore backward-looking, does not provide the framework needed to guide EPA's efforts into the future. We recognize that many of the challenges that EPA faces are rooted in TSCA but continue to believe that, without a plan that incorporates leading strategic planning practices—particularly a plan that clearly articulates how EPA will address management challenges—EPA cannot be assured that it its new approach to managing chemicals will provide a framework to effectively guide its efforts. Consequently, EPA could be investing valuable resources, time, and effort without being certain that its investments will bring the agency closer to achieving its goal of ensuring the safety of chemicals.

David C. Trimble
Director, Natural Resources and Environment

Appendix I. Scope and Methodology

This report focuses on the Environmental Protection Agency's (EPA) management of chemicals within the limits of its existing authorities under the Toxic Substances Control Act (TSCA). Our objectives were to determine the extent to which (1) EPA has made progress implementing its new approach to managing toxic chemicals under its existing TSCA authority, and (2) EPA's new approach positions the agency to achieve its goal of ensuring the safety of chemicals.

To determine the extent to which EPA has made progress implementing its new approach to managing toxic chemicals under its existing TSCA authority, we focused on EPA's new initiatives to manage chemicals beginning in September 2009.This was the date that EPA's Administrator announced that the agency was pursuing a comprehensive approach to enhance EPA's current chemicals management program within the limits of existing authorities. We identified and reviewed documents associated with activities that EPA identified as being part of its new approach, including the agency's Chemical Action Plans for 10 chemicals or chemical categories, its February 2012 *Existing Chemicals Program Strategy,* and March 2012 TSCA Work Plan and associated TSCA Work Plan Methods document. We also reviewed documents, including *Federal Register* Notices, and interviewed agency officials, with respect to EPA's efforts to obtain test data for High Production Volume Challenge chemicals. We reviewed applicable legislation, including the relevant sections of TSCA; relevant federal regulations; and EPA's policies and procedures on EPA's existing and new chemical programs. We analyzed TSCA rulemaking trends from January 2001 through October 2012. We selected this time frame to include a time period prior to and after the announcement of EPA's new approach to managing chemicals to see how rulemaking changed over this time period. In our report, we only present findings from 2009 to 2012 when the majority of the rulemaking activity took place. To identify TSCA rules that were issued, we searched online sources, including *Federal Register* Notices and the information available from the Office of Management and Budget (OMB), Office of Information and Regulatory Affairs (OIRA). We then analyzed proposed, final, and direct final rules exclusive of rule revocations, technical modifications, and withdrawals for rulemakings deriving from the implementation of Title I of TSCA. We excluded rules deriving from the implementation of other titles of TSCA from this rulemaking analysis as they derive from separate provisions or later amendments to TSCA; they are chemical-specific provisions that have specific rulemaking requirements and deadlines. From the list of identified TSCA rules, we determined frequencies of TSCA rulemaking by TSCA section; for example, we determined the number of proposed and final significant new use rules under TSCA section 5, as well as the number of chemicals subject to these rules. We also analyzed such frequencies before and after 2009, which was when EPA announced its new approach to managing chemicals. As part of our rulemaking analysis, we also examined EPA's Action Development

Process (ADP) and interviewed officials from EPA's Office of Policy, Office of Chemical Safety and Pollution Prevention (OCSPP), and Office of Pollution Prevention and Toxics (OPPT) on how individual TSCA rules are developed. We obtained and reviewed EPA's regulatory plans and agendas from 2011, to determine if TSCA rules were expected to be issued in proposed or final form during this period. We also reviewed Executive Order 12866 to characterize EPA's and OMB's processes for submitting and reviewing TSCA rules, respectively. In addition, we reviewed EPA's efforts, through discussions with EPA officials, and by reviewing recently issued policies, on confidential business information claims.

To determine the extent to which EPA's new approach positions the agency to achieve its goal of ensuring the safety of chemicals, we reviewed EPA's *Existing Chemical Program Strategy* , which EPA identified as the document that it intends to use to guide its current and future efforts, and compared it against leading practices in federal strategic planning, which include practices such as developing strategies to address challenges that may threaten the agency ability to achieve its goal including identifying the resources needed to achieve agency goals.[47] To further characterize how EPA is positioning itself to assess and control chemicals, we asked EPA officials to describe expected outcomes from EPA's recent rulemakings, for example, the extent to which TSCA rules would position EPA to obtain chemical data and reduce chemical risks. We also obtained documents and interviewed EPA officials on the agency's processes for screening existing chemicals to identify those requiring further review or a risk assessment. We also requested and obtained information on EPA's resource levels to carry out this work—particularly related to the resources required to conduct risk assessments indentified in its TSCA Work Plan. We reviewed EPA estimates on the costs and staff levels for performing such risk assessments.

For both objectives, we interviewed officials with EPA's Office of Chemical Safety and Pollution Prevention, including its Office of Pollution Prevention and Toxics (OPPT), the office with primary responsibility for implementing TSCA, regarding EPA's efforts to manage chemicals. Specifically, we interviewed officials across OPPT's divisions about how EPA obtains chemical data, analyzes existing chemicals, and conducts risk assessments, as well as agency reviews of new chemicals and new uses of existing chemicals, reviews of claims of confidential business information, and the process of promulgating rules under TSCA. We also interviewed officials from EPA's Office of Research and Development relating to EPA's development and use of new analytical methods and tools. In addition, we interviewed representatives from industry, including the American Chemistry Council (a national chemical manufacturers association); the Society of Chemical Manufacturers and Affiliates (a national, specialty chemical manufacturers association); American Fuel & Petrochemical Manufacturers (a trade association representing high-tech American fuel manufacturers); Consumer Specialty Products Association (a trade association representing companies manufacturing various cleaning products); Bergeson & Campbell, P.C. (a law firm that represents chemical manufacturers); and Greenwood Environmental Counsel (a law firm). We also interviewed other stakeholder groups, including the Environmental Defense Fund (a national, nonprofit environmental advocacy organization); Safer Chemicals, Healthy Families (a coalition of individuals, health professionals, advocates for people with learning and developmental disabilities, reproductive health advocates, environmentalists and businesses); and the Environmental Law Institute (a nonpartisan research and education center working to strengthen environmental protection by improving law and governance worldwide. We selected these stakeholder groups based on discussions with other stakeholders and our prior work on TSCA.[48] We sought to achieve a balance of perspectives between groups representing industry and those representing the environment.

To assess the reliability of EPA's data related to the types and numbers of TSCA rules promulgated, data from EPA on rules under OMB review over 90 days, and data on the number of chemicals regulated under TSCA and the number of chemicals under assessment, we reviewed relevant documents and interviewed knowledgeable agency officials. We worked with the EPA to ensure that we had the most updated data and, in consultation with the EPA officials, revised the numbers of chemicals listed for commercial use under TSCA and the numbers of filings to reflect the most updated numbers. Based on this review and our discussions with the EPA officials, we concluded that the data were sufficiently reliable for the purposes of reporting on rulemaking trends and the numbers of chemicals listed with EPA for commercial use as authorized under TSCA.

We conducted this performance audit from December 2011 to March 2013 in accordance with generally accepted government auditing standards. Those standards require that we plan and perform the audit to obtain sufficient, appropriate evidence to provide a reasonable basis for our findings and conclusions based on our audit objectives. We believe that the evidence obtained provides a reasonable basis for our findings and conclusions based on our audit objectives.

End Notes

[1] Processing refers to the preparation of a chemical substance or mixture, after its manufacture, for distribution in commerce.

[2] Excluded substances include certain nuclear materials, pesticides, food, food additives, tobacco, drugs, and cosmetics.

[3] Throughout this report, we use the phrase "chemical companies" to refer generally to companies that manufacture, import, process, distribute in commerce, use, or dispose of chemicals regulated under TSCA. When it is important to differentiate between, for example, manufacturers and processors, we specify the type of company to which we are referring.

[4] Toxicity represents the degree to which a chemical is harmful. In this report, the terms toxicity and hazard are used synonymously. Exposure represents the magnitude, frequency, and duration of contact with a chemical.

[5] GAO, *Toxic Substances Control Act: Legislative Changes Could Make the Act More Effective,* GAO/RCED-94-103 *(Washington, D.C.: Sept. 26, 1994), GAO, Chemical Regulation: Options Exist to Improve EPA's Ability to Assess Health Risks and Manage Its Chemical Review Program*, GAO-05-458 (Washington, D.C.: June 13, 2005).

[6] GAO-05-458.

[7] *GAO, High-Risk Series: An Update.* GAO-09-271 (*Washington, D.C.: Jan. 22, 2009).*

[8] We have reported on EPA's difficulty producing timely, credible IRIS assessments, which contributed to our decision in 2009 to add EPA's processes for assessing and controlling chemicals to our list of areas at high risk for waste, fraud, abuse, and mismanagement or in need of broad-based transformation.

[9] GAO-05-458.

[10] Companies are not required to submit new notices if any information provided in the original notice changes.

[11] GAO-05-458.

[12] Of the over 84,000 chemicals currently on the TSCA inventory, approximately 8,000 chemicals are produced at annual volumes of 25,000 pounds or greater. TSCA requires EPA to compile, keep current, and publish a list of each chemical substance that is manufactured or processed in the United States.

[13] EPA came to this conclusion after reviewing over 100 studies of the health risks of asbestos, as well as public comments on the proposed rule.

[14] Existing chemicals listed on the TSCA inventory are composed of approximately 62,000 that were in commerce in 1979 when EPA began reviewing chemicals and 22,000 chemicals that were listed for commercial use after 1979. Under EPA's periodic chemical data reporting rules, approximately 8,000 chemicals are reported as being produced or imported at annual volumes of 25,000 pounds or greater but, for the remaining 76,000 chemicals on the inventory, EPA does not know how many are currently in commerce.

[15] Final rules are located at 40 C.F.R. §§ 799.5087 and 799.5089 (2012). The proposed rule is located at 76 Fed. Reg. 65580 (Oct. 21, 2011).

[16] In addition to promulgating test rules, EPA told us that it has required testing for an additional 68 chemicals in enforceable consent agreements.

[17] Under the 1998 High Production Volume (HPV) challenge program, EPA asked chemical companies to sponsor chemicals that were manufactured in or imported into the United States in quantities greater than 1 million pounds per year—at the time, 2,800 chemicals—and voluntarily provide existing toxicity data and, if data did not exist, conduct testing and provide EPA with the resulting data. Companies sponsored more than 2,200 HPV chemicals.

[18] The data set sought by EPA is known as the Screening Information Data Set, and was developed by the Organization for Economic Cooperation and Development. The data set provides an internationally agreed upon set of test data for screening high-production

volume chemicals for human and environmental hazards, and it is intended to allow the EPA and others to make an informed, preliminary judgment about the hazards of HPV chemicals. The data set is not intended to describe a chemical thoroughly, but rather it is intended to provide enough information to support an initial (or screening level) assessment and to assign a priority for further work, if necessary.

[19] For example, the production threshold for providing processing and use information went from 300,000 pounds or more to 100,000 pounds or more in 2012 and will be reduced to 25,000 pounds thereafter.

[20] The European Chemicals Agency implements the European Union's chemicals legislation. The European Union's chemicals legislation requires companies to develop information on chemicals' effects on human health and the environment before entering commerce, while TSCA does not require companies to develop such information absent EPA rulemaking requiring them to do so.

[21] EPA identifies a broad range of activities that may cause a person or entity to be considered a chemical processor. Examples of chemical processors include, but are not limited to, producers of paint, automotive products, specialty cleaners, rubber, and plastics; or tire manufacturers, tanneries, textile mills, and metal coating facilities.

[22] Section 11(c) of TSCA gives EPA the authority, in carrying out the provisions of TSCA, to use a subpoena to require the attendance and testimony of witnesses and the production of reports, papers, documents, answers to questions, and other information that the EPA Administrator deems necessary.

[23] In 2011, EPA convened a stakeholder meeting to discuss proposed screening criteria and data sources and took public comment over a 35-day period. Based on the input received, EPA devised and executed a protocol that used a combination of risk factors and other criteria. Using this protocol, EPA winnowed an initial group of 1,235 chemicals down to 83.

[24] These chemicals are: antimony and antimony compounds, HHCB (1,3,4,6,7,8- hexahydro-4,6,6,7,8,8,-hexamethylcyclopenta[g]-2-benzopyran), long-chain chlorinated paraffins, medium-chain chlorinated paraffins, methylene chloride, n-methylpyrrolidone, and trichloroethylene.

[25] Prior to its effort under its TSCA Work Plan, EPA conducted two TSCA-related risk assessments since 2001.

[26] 78 Fed. Reg. 1856 (Jan. 9, 2013).

[27] Antimony trioxide (ATO) as a synergist in halogenated flame retardants; and 1,3,4,6,7,8-Hexahydro-4,6,6,7,8,8,-hexamethylcyclopenta-[γ]-2-benzopyran (HHCB) as a fragrance ingredient in commercial and consumer products.

[28] Methylene chloride or dichloromethane (DCM) and n-methylpyrrolidone (NMP) in paint stripper products.

[29] Trichloroethylene (TCE) as a degreaser and a spray-on protective coating.

[30] Medium and long chain chlorinated paraffins.

[31] According to EPA officials, the IRIS program is charged with evaluating chronic hazard potential including hazard identification and dose response information for cancer and noncancer outcomes. Many of the chemical uses and particularly the consumer uses are acute and short term in nature. In most cases, the information used to develop the dose-response assessments is based on intermittent exposures to workers or animals in a controlled environment. IRIS assessments include an adjustment to continuous exposure in the derivation of toxicity values. Often data are available in IRIS for shorter term exposures scenarios that have long-lasting/ persistent effects (e.g., development toxicity). In these cases, evaluation of hazard and dose response information described in an IRIS assessment is useful.

[32] GAO-08-440.

[33] This focus is consistent with the fact that media-specific environmental laws such as the Clean Air Act and Clean Water Act are available to limit the concentration of contaminants in water or the ambient air, and TSCA generally requires EPA to defer action to such other

laws. However, information on such continuous exposures is still critical for regulation under TSCA. For example, to promulgate a rule under section 6 of TSCA, EPA must establish the effects of a chemical on health and the environment and the magnitude of the exposure of human beings and the environment to such a chemical.

[34] EPA Office of Research and Development, *Human Health Risk Assessment Strategic Research Action Plan 2012-2016* (June 2012), 15.

[35] Any rules that EPA plans to issue under TSCA that are considered significant regulatory actions, as defined by Executive Order 12866, are subject to review by the Office of Information and Regulatory Affairs, an office within OMB, prior to being proposed in the *Federal Register*. Among other things, a significant regulatory action may have an annual effect on the economy of $100 million or more or raise novel legal or policy issues.

[36] These three groups are: (1) a category of eight phthalates, (2) a category of polybrominated diphenyl ether (PBDE), and (3) bisphenol A (BPA).

[37] Under Executive Order 12866, the review period may be extended by the head of the rulemaking agency, and the OMB Director may extend the review period once for no more than 30 days.

[38] In spring 2012, EPA proposed three significant new use rules that would require companies to report new uses of five groups of chemicals, including in domestic and imported articles.

[39] EPA has used this approach before but infrequently. EPA first eliminated the article exemption for a chemical substance in 1991, when it promulgated a significant new use rule for erionite fiber, and it used the same approach for a significant new use rule pertaining to the use of elemental mercury in certain switches in 2007.

[40] Chemical identity is a name that uniquely identifies a chemical. This can be a name in accordance with the nomenclature systems of, for example, the International Union of Pure and Applied Chemistry (IUPAC) or the Chemical Abstracts Service (CAS).

[41] EPA determined that many of the identified health and safety filings (1) were misidentified as having confidentiality claims for the chemical name or as being or containing health and safety studies, (2) had valid confidentiality claims for the chemical name, or (3) contained claims that were considered invalid under the law.

[42] The first policy, published in January 2010, involves confidentiality claims related to chemical identities included in health and safety studies submitted under section 8(e) of TSCA. Under this policy, when the agency receives a health and safety study for a chemical already listed on the public portion of the TSCA inventory, it will not accept a claim that the identity of the chemical is confidential. The second policy, published in May 2010, involves confidentiality claims related to chemical identities and data included in health and safety studies more generally. Under this policy, EPA does not extend confidential treatment to health and safety studies or data contained in health and safety studies unless the studies or data reveal (1) processes used in making the chemical or (2) the portion of the chemical in a mixture.

[43] In August 2012, EPA provided us with an estimate that stated that it needed 17.5 full- time EPA staff and about $800,000 in contractor support to prepare the 7 risk assessments; it needed an additional $160,000 per chemical to support contractor peer reviews. According to EPA officials, they will be in a better position to provide an accounting of expenditures toward the end of fiscal year 2013.

[44] GAO, *Environmental Justice: EPA Needs to Take Additional Actions to Help Ensure Effective Implementation*, GAO-12-77 (Washington, D.C.: Oct. 6, 2011); GAO, *Environmental Protection: EPA Should Develop a Strategic Plan for Its New Compliance Initiative*, GAO-13-115 (Washington, D.C.: Dec. 10, 2012).

[45] Leading practices in federal strategic planning include defining mission and goals, involving leadership and stakeholders, developing performance measures, and developing strategies to address management challenges and resources needed, among others.

[46] The strategic planning elements established under the Government Performance and Results Act (GPRA) of 1993 and associated OMB guidance and practices we identified, taken

together, can serve as leading practices for strategic planning at lower levels within federal agencies, such as planning for individual divisions, programs, or initiatives. For example, see GAO, *Executive Guide: Effectively Implementing the Government Performance and Results Act*, GAO/GGD-96-118 (Washington, D.C.: June 1, 1996); GAO, *Tax Administration: IRS Needs to Further Refine Its Tax Filing Season Performance Measures*, GAO-03-143 (Washington, D.C.: Nov. 22, 2002); and GAO, *Managing for Results: Strengthening Regulatory Agencies' Performance Management Practices*, GAO/GGD-00-10 (Washington, D.C.: Oct. 28, 1999).

[47] GAO-12-77 and GAO-13-115.

[48] GAO-05-458 and GAO, *EPA Should Focus Its Chemical Use Inventory on Suspected Harmful Substances*, GAO/RCED-95-165 (Washington, D.C.: July 7, 1995).

In: Environmental Protection Agency ... ISBN: 978-1-62808-244-9
Editor: Robin Laurits

Chapter 2

THE TOXIC SUBSTANCES CONTROL ACT (TSCA): A SUMMARY OF THE ACT AND ITS MAJOR REQUIREMENTS*

Linda-Jo Schierow

SUMMARY

This report summarizes the Toxic Substances Control Act (TSCA) and the major regulatory programs dealing with chemical production and distribution in U.S. commerce.

The President's Council on Environmental Quality proposed comprehensive federal legislation in 1971 to identify and control potentially dangerous chemicals in U.S. commerce that were not adequately regulated under other environmental statutes. President Ford signed TSCA into law on October 11, 1976. Subsequently, five titles were added to address specific concerns—asbestos in 1986 (Title II, P.L. 99-519), radon in 1988 (Title III, P.L. 100-551), lead in 1992 (Title IV, P.L. 102-550), environmental and energy issues in schools in 2007 (Title V, P.L. 110-140), and formaldehyde emissions from composite wood products in 2010 (Title VI, P.L. 111-199). In 2008, Congress added provisions to Title I, Section 6 and Section 12, banning certain activities with respect to elemental mercury (P.L. 110-414).

* This is an edited, reformatted and augmented version of a Congressional Research Service publication, CRS Report for Congress RL31905, prepared for Members and Committees of Congress, from www.crs.gov, dated April 1, 2013.

TSCA authorizes EPA to identify potentially dangerous chemicals in U.S. commerce that should be subject to federal control. The act authorizes EPA to gather and disseminate information about production, use, and possible adverse effects to human health and the environment of existing chemicals, and to issue "test rules" that require manufacturers and processors of potentially dangerous chemicals to conduct and report the results of scientific studies to fill information gaps. For chemicals new to U.S. commerce, TSCA requires pre-market screening and regulatory tracking of new chemical products.

If EPA identifies unreasonable risks associated with existing or new chemicals, TSCA requires the agency to initiate rulemaking to reduce risks to a reasonable level. EPA may regulate the manufacture, importation, processing, distribution, use, and/or disposal of chemicals. TSCA provides a variety of regulatory tools to EPA, ranging in severity from a total ban on production, import, and use to a requirement that a product must bear a warning label at the point of sale. However, TSCA directs EPA to use the least burdensome option that can reduce risk to a level that is reasonable, given the benefits provided by the chemical product or process.

Title I of the original statute establishes the core program, directs EPA to control risks from polychlorinated biphenyls (PCBs), and bans certain activities with respect to elemental mercury. Title II directs EPA to set standards for asbestos mitigation in schools and requires asbestos contractors to be trained and certified. Title III directs EPA to provide technical assistance to states that choose to support radon monitoring and control. Title IV provides similar assistance with respect to abatement of lead-based paint hazards. Title V addresses environmental issues at schools, including energy efficiency. Finally, Title VI directs EPA to set standards for emissions of formaldehyde from composite wood products.

INTRODUCTION

The Environmental Protection Agency (EPA) is responsible for identifying and regulating toxic substances in U.S. commerce under the authority of the Toxic Substances Control Act (TSCA). This report defines key terms, provides a brief history of toxic substances control law, and describes key provisions of TSCA. In addition, this report lists several references for more detailed information about the act and provides a table that cross references sections of the *U.S. Code* with corresponding sections of the act. The report is descriptive rather than analytic, highlights key provisions rather than providing a comprehensive inventory of the act's numerous sections, and addresses authorities and limitations imposed by statute, rather than the status of EPA implementation or other policy issues.

OVERVIEW

The Toxic Substances Control Act (15 U.S.C. 2601 *et seq.*) authorizes the EPA to screen existing and new chemicals used in manufacturing and commerce to identify potentially dangerous products or uses that should be subject to federal control. Both naturally occurring and synthetic chemicals are subject to TSCA, with the exception of chemicals regulated under other federal laws concerning food, drugs, cosmetics, firearms, ammunition, pesticides, tobacco, or mixtures. EPA may require manufacturers and processors of chemicals to conduct and report the results of tests to determine the effects of potentially dangerous chemicals on living things. Based on test results and other information, EPA must regulate the manufacture, importation, processing, distribution, use, and/or disposal of any chemical that presents an unreasonable risk of injury to human health or the environment. A variety of regulatory tools is available to EPA under TSCA, ranging in severity from a total ban on production, import, and use to a requirement that a product bears a warning label at the point of sale. TSCA directs EPA to use the least burdensome option that can reduce risk to a level that is reasonable, given the benefits provided by the chemical product or process.

Table 1. Toxic Substances Control Act and Major Amendments (codified as 15 U.S.C. 2601-2671)

Year	Act	Public Law Number
1976	Toxic Substances Control Act	P.L. 94-469
1986	Asbestos Hazard Emergency Response Act	P.L. 99-519
1988	Radon Program Development Act	P.L. 100-551
1990	Radon Measurement	P.L. 101-508, §10202
1990	Asbestos School Hazard Abatement Reauthorization Act	P.L. 101-637
1992	Residential Lead-Based Paint Hazard Reduction Act of 1992	P.L. 102-550
2007	Energy Independence and Security Act of 2007, subtitle E - Healthy High-Performance Schools	P.L. 110-140
2008	Mercury Export Ban Act of 2008	P.L. 110-414
2010	Formaldehyde Standards for Composite Wood Products Act	P.L. 111-199

Federal legislation to control toxic substances was originally proposed in 1971 by the President's Council on Environmental Quality. Its report, "Toxic Substances," defined a need for comprehensive legislation to identify and control chemicals whose manufacture, processing, distribution, use, and/or disposal was potentially dangerous and not adequately regulated under other environmental statutes. The House and Senate each passed bills in both the 92nd and 93rd Congresses (in 1972 and 1973, respectively), but controversies over the scope of chemical screening prior to commercial production and distribution, level of costs, and the relationship to other regulatory laws stalled final action. Episodes of environmental contamination—including contamination of the Hudson River and other waterways by PCBs, the threat of stratospheric ozone depletion from chlorofluorocarbon (CFC) emissions, and contamination of agricultural produce by polybrominated biphenyls (PBBs) in the state of Michigan—together with more exact estimates of the costs of imposing toxic substances controls, opened the way for final passage of the legislation. President Ford signed the TSCA into law on October 11, 1976. The original legislation included a single title, which has since been designated Title I.

TSCA (Title I) directs EPA to

- require manufacturers and processors to conduct tests for existing chemicals if (1) their manufacture, distribution, processing, use, or disposal may present an unreasonable risk of injury to health or the environment; or they are to be produced in substantial quantities and the potential for environmental release or human exposure is substantial or significant; (2) existing data are insufficient to predict the effects of human exposure and environmental releases; and (3) testing is necessary to develop such data (§4);
- prevent future risks through pre-manufacture screening and regulatory tracking of new chemical products (§5);
- control unreasonable risks already known, including risks from polychlorinated biphenyls (PCBs), as well as risks for existing chemicals that may be discovered in the future (§6); and
- gather and disseminate information about chemical production, use, and possible adverse effects to human health and the environment (§8).

Authorization for appropriations for these activities and a state grant program for control of toxic substances in the environment expired on

September 30, 1983, although appropriations for these programs have continued.

In October 2008, Congress amended TSCA Title I when it enacted the Mercury Export Ban Act of 2008 (P.L. 110-414). It prohibits certain activities with respect to elemental mercury.

Subsequently, five titles have been added to TSCA to address specific concerns—asbestos in 1986 (Title II, P.L. 99-519), radon in 1988 (Title III, P.L. 100-551), lead in 1992 (Title IV, P.L. 102-550), schools in 2007 (Title V, P.L. 110-140), and formaldehyde in 2010 (Title VI, P.L. 111- 199). Title II directs EPA to set standards for asbestos mitigation in schools, and requires asbestos contractors to be trained and certified. Title III directs EPA to provide technical assistance to states that choose to support radon monitoring and control. Title IV provides similar assistance with respect to abatement of lead-based paint hazards. Title V addresses environmental issues at schools, including energy efficiency. Finally, Title VI establishes limits on emissions of formaldehyde from composite wood products.

TITLE I

Testing of Chemicals

Many chemicals, even some in widespread use, are not well characterized in terms of their potential health and environmental effects. One of the major goals of TSCA was to induce the development of test data by producers (i.e., manufacturers, importers, and processors) of chemicals in commerce. Section 4 of TSCA directs EPA to require the development of test data on existing chemicals when certain conditions prevail: (1) the manufacture, processing, distribution, use, or disposal of the chemical "may present an unreasonable risk;" or (2) the chemical is produced in very large volume and there is a potential for a substantial quantity to be released into the environment or for substantial or significant human exposure. Under either condition, EPA must issue a rule requiring tests if (a) existing data are insufficient to resolve the question of safety, and (b) testing is necessary to develop the data.

Because there were more than 55,000 chemicals in U.S. commerce at the time EPA was to begin developing test rules, Congress established a special interagency committee to help EPA determine which chemicals should be considered first, and to coordinate testing needs and efforts among government agencies. At least every six months the Interagency Testing Committee (ITC) must consider candidate chemicals for inclusion on a list of substances that the ITC recommends to EPA for development and promulgation of test rules. TSCA directs the ITC to "designate" a subset of chemicals on the list for EPA action within 12 months. The list can contain no more than 50 "designated" chemicals at any time. When a chemical is designated, EPA has one year to respond by issuing a proposed test rule or a notice explaining why no testing is needed.

TSCA requires the ITC to consider the following factors when it makes listing decisions: (1) quantity of the substance to be manufactured, (2) quantity of the chemical in environmental releases, (3) number of people who will be exposed occupationally and the duration of exposure, (4) extent of non-occupational human exposure, (5) similarity of the chemical to any other chemical known to present an unreasonable risk, (6) existence of data concerning environmental or health effects of the chemical, (7) the quantity of information to be gained by testing, and (8) the availability of facilities and personnel for performing testing. Chemicals known or suspected to cause or contribute to cancer, gene mutations, or birth defects are to be assigned a higher priority. In response to information that indicates "there may be a reasonable basis to conclude that a chemical ... presents or will present a significant risk of serious or widespread harm to human beings from cancer, gene mutations, or birth defects," TSCA requires EPA action to prevent or reduce that risk or publication of a finding that the risk is not unreasonable.

Pre-manufacture Notification for New Chemicals or Uses

TSCA (§5) requires manufacturers, importers, and processors to notify EPA at least 90 days prior to producing or otherwise introducing a new chemical product into the United States. Any information or test data that is known to, reasonably ascertainable by, or in possession of the notifier, and that might be useful to EPA in evaluating the chemical's potential adverse effects on human health or the environment, must be submitted to EPA at the same time. TSCA also requires EPA to be notified when there are plans to produce, process, or use an existing chemical in a way that differs from previously permitted uses, if the Administrator has determined by rule that new uses of the chemical may produce significant changes in human and environmental exposures and therefore require notification. The 90-day notice provides EPA with the opportunity to evaluate the chemical use and, if necessary, to prohibit or limit such activity before it occurs to prevent unreasonable risk of injury to human health or the environment.

EPA has 45 days after notification (or up to 90 days if it extends the period for good cause) to evaluate the potential risk posed by the chemical. If EPA determines that there is a reasonable basis to conclude that the substance presents or will present an unreasonable risk, the Administrator must promulgate requirements to protect adequately against such risk. Alternatively, EPA may determine that the proposed activity related to a chemical does not present an unreasonable risk; this decision may be based on the available data, or, when no data exist to document the effects of exposure, on what is known about the effects of chemicals in commerce with similar chemical structures and used in similar ways.

The purpose of EPA's screening procedure is to identify potential hazards, and control them before use of a chemical becomes widespread. If data are inadequate to make an informed judgment and (1) manufacture, processing, distribution in commerce, use, or disposal may present an unreasonable risk; or (2) a chemical is to be produced in substantial quantities, and the potential for environmental release or human exposure is substantial or significant, EPA may issue a proposed order to prohibit or limit such activities until sufficient data are submitted.

Although the legislative history of TSCA includes a presumption that testing of new products would take place before they were widely used, either as the chemical was developed, or as its markets grew, TSCA also forbids promulgation of blanket testing requirements for all new chemicals. This reflects concern that uniform testing requirements might stifle innovation in the chemical industry. Thus, EPA must decide which chemicals, or which categories of chemicals, warrant the costs of pre-market testing. EPA reviews approximately 1,000 new chemical manufacturing notices annually.

Regulatory Controls for Hazardous Chemicals

TSCA requires EPA to regulate manufacturing, processing, distribution in commerce, use, or disposal of a chemical if it will present an unreasonable risk of injury to health or the environment, and the risk cannot be reduced to a sufficient degree under another federal law administered by EPA. The alternative means available to EPA for controlling chemical hazards that present unreasonable risks are specified in Section 6 of TSCA. EPA has the authority to

- prohibit or limit the amount of production or distribution of a substance in commerce;
- prohibit or limit the production or distribution of a substance for a particular use;
- limit the volume or concentration of the chemical produced;
- prohibit or regulate the manner or method of commercial use;
- require warning labels and/or instructions on containers or products;
- require notification of the risk of injury to distributors and, to the extent possible, consumers;
- require record-keeping by producers;
- specify disposal methods; and
- require replacement or repurchase of products already distributed.

EPA also may impose any of these requirements in combination or for a specific geographical region. However, EPA is required by TSCA to regulate only "to the extent necessary to protect adequately" against a risk, and to use the "least burdensome" regulatory approach, even in controlling unreasonable risks.

Two chemical substances are directly addressed in Title I: PCBs and elemental mercury. TSCA directs EPA to regulate PCBs and to ban most uses. In addition, TSCA prohibits the sale, distribution, or transfer of elemental mercury by federal agencies.

Information Gathering

Section 8 of TSCA requires EPA to develop and maintain an inventory of all chemicals, or categories of chemicals, manufactured or processed in the United States. The first version of this inventory identified approximately 55,000 chemicals in commerce in 1979. All chemicals not on the inventory are, by definition, "new" and subject to the notification provisions of Section 5. These chemicals must be added to the inventory if they enter U.S. commerce. Chemicals need not be listed if they are only produced in very small quantities for purposes of experimentation or research.

To aid EPA in its duties under TSCA, the agency was granted considerable authority to collect information from industries. EPA may require maintenance of records and reporting of chemical identities, names, and molecular structures; categories of use; amounts manufactured and processed for each category of use; descriptions of byproducts resulting from manufacture, processing, use, and disposal; environmental and health effects; number of individuals exposed; number of employees exposed and the duration of exposure; and manner or method of chemical disposal.

Manufacturers, processors, and distributors of chemicals are required to maintain records of significant adverse reactions to health or the environment alleged to have been caused by a substance or mixture. Records of adverse effects on the health of employees must be retained for 30 years from the date of reporting. Industry also must submit lists and copies of health and safety studies. Studies showing adverse effects previously unknown must be submitted to EPA as soon as they are completed or discovered.

Imminent Hazards

Section 7 provides EPA authority to take emergency action through the district courts to control a chemical substance or mixture which presents an imminent and unreasonable risk of serious widespread injury to health or the environment.

Relation to Other Laws

Section 9 requires EPA to report cases of chemical risk to other federal agencies with the authority to prevent or reduce the risk. If the other agency determines that the activity does not present the risk described by EPA in its report, or initiates action under the law to protect against the risk, TSCA prohibits EPA from taking regulatory action with respect to such risk. For statutes under EPA's jurisdiction, TSCA gives the Administrator discretion to decide if a risk can best be handled under the authority of TSCA.

Enforcement and Judicial Review

Section 11 authorizes EPA to inspect any facilities subject to TSCA requirements and to issue subpoenas requiring attendance and testimony of witnesses, production of reports and documents, answers to questions and other necessary information. Section 13 mandates TSCA enforcement at the national borders by the Treasury Department.

Section 15 identifies acts prohibited under TSCA, while Section 16 describes penalties for acts violating these prohibitions, as well as recourse available to anyone accused of such violations. Section 16 authorizes civil penalties, not to exceed $25,000 per violation per day, and affords the defendant an opportunity to request a hearing before an order is issued and to petition for judicial review of an order after it is issued. Criminal penalties also are authorized for willful violations. Section 17 provides jurisdiction to U.S. district courts in civil actions to enforce TSCA Section 15 by restraining or compelling actions that violate or comply with it, respectively. Chemicals may be seized and condemned if their manufacture, processing, or distribution violated the act. Section 19 authorizes any person to file a petition for judicial review of specified rules within 60 days of issuance under TSCA. The court is directed to set aside specified rules if they are not supported by substantial evidence in the rulemaking record taken as a whole.

Section 20 authorizes civil suits by any person against any person in violation of the act. It also authorizes suits against EPA to compel performance of nondiscretionary actions under TSCA. Section 21 provides the public with the right to petition for the issuance, amendment, or repeal of a rule requiring toxicity testing of a chemical, regulation of the chemical, or reporting.

Confidential Business Information

Section 14 provides broad protection of proprietary confidential information about chemicals in commerce. Disclosure by EPA employees of such information generally is not permitted, except to other federal employees, or when necessary to protect health or the environment. Data from health and safety studies of chemicals are not protected unless their disclosure would reveal a chemical process or chemical proportion in a mixture. Wrongful disclosure of confidential data by federal employees is prohibited, and may result in criminal penalties.

Chemical Categories

Section 26 allows EPA to impose regulatory controls on categories of chemicals, rather than on a case-by-case basis. However, EPA cannot regulate a group merely because it is composed of new chemical substances.

State Preemption

TSCA Section 18 preempts state actions that establish or continue in effect requirements applicable to a chemical substance or mixture that is regulated under TSCA Section 5 or 6, unless the state requirement is identical to the federal requirement, implements another federal law, or prohibits use of the substance or mixture within the state. However, a state may ask EPA to allow a state requirement that provides a significantly higher degree of protection from risk than does the federal requirement.

Other Provisions

TSCA Section 10 directs EPA to conduct and coordinate among federal agencies research, development, and monitoring that is necessary to the purposes of the act.

Section 12 excludes chemical products manufactured for export from TSCA requirements except for reporting and record keeping requirements in Section 8. In 2008, Congress excluded elemental mercury from this exemption, banning its export beginning in 2013, with the exception of mercury contained in coal. Other exemptions for essential uses may be granted by rule.

Section 22 waives compliance when in the interest of national defense.

Section 23 provides protection of employees who assist in carrying out the provisions of the act (i.e., "whistle-blowers").

The potential effects of TSCA rules on employment must be monitored by EPA, according to Section 24.

Section 25 mandates study of the need for indemnification of people affected by federal laws administered by EPA and of the feasibility of establishing a standard classification system for chemical substances and of storing and retrieving information about them.

Section 26 authorizes data sharing and cooperative action to facilitate TSCA implementation between EPA and other federal agencies. It also authorizes collection of fees for EPA processing of data submitted in response to an order under Section 4 or 5. EPA is directed to establish an office to assist the regulated community. The agency also must establish a procedure to ensure disclosure of financial interests in the regulated community by EPA employees. Final orders issued under TSCA must contain a statement of basis and purpose. Finally, Section 26 established within EPA a new Assistant Administrator for Toxic Substances.

TSCA Section 27 authorizes research and development of test methods for chemicals by the Public Health Service in cooperation with EPA.

Grants to states are authorized by Section 28 to establish and operate programs to prevent or eliminate unreasonable risks to health or the environment.

Section 29 authorized appropriations through 1983. An annual report is mandated by Section 30.

TITLE II (ASBESTOS IN BUILDINGS)

Growing public concern about the presence of potentially hazardous asbestos in buildings, especially in schools, led to congressional efforts to address this problem. Title II of TSCA, the Asbestos Hazard Emergency Response Act (AHERA), was enacted in 1986 (P.L. 99-519) and amended in July 1988 (P.L. 100-368). It required EPA to set standards by October 1987, for responding to the presence of asbestos in schools. The standards, set at levels adequate to protect public health and the environment, identify appropriate response actions that depend on the physical condition of asbestos. Schools, in turn, were required to inspect for asbestos-containing material, and to develop and implement a plan for managing any such material. Plans for managing asbestos were to be submitted by schools before May 1989, and implementation was to begin by July 1989. The law contains no deadlines for schools to complete implementation.

Title II requires asbestos contractors and analytical laboratories to be certified, and schools to use certified persons for abatement work. Training and accreditation requirements also apply to inspectors, contractors, and workers performing asbestos abatement work in all public and commercial buildings. EPA may award training grants to nonprofit organizations for asbestos health and safety programs. However, authorization of appropriations for this grant program expired September 30, 1995. Other Title II requirements (such as mandates that buildings be inspected for asbestos) have not been extended to non-school buildings.

To enforce requirements, TSCA authorizes EPA to take emergency action with respect to schools if school officials do not act to protect children. The act also authorizes citizen action with respect to asbestos-containing material in a school and to compel action by EPA, either through administrative petition or judicial action. Civil penalties not to exceed $5,000 are authorized for violations such as failing to conduct an inspection or to develop a school management plan.

Concern about how schools would pay for required actions was addressed in separate legislation (the Asbestos School Hazard Abatement Act of 1984, or ASHAA, P.L. 98-377). It established a program offering grants and interest-free loans to schools with serious asbestos problems and demonstrated financial need. Although EPA for several years did not request funding for this program, Congress appropriated funds. Authorization of appropriations for this program expired September 30, 1995, and Congress has not appropriated funds since FY1993; a total of $382 million in grant and loan funds were appropriated from FY1984 through FY1993. Repaid ASHAA loans are returned to an Asbestos Trust Fund, established in TSCA Title II, to become a dedicated source of revenues for future asbestos control projects.

TITLE III (RADON PROGRAMS)

In October 1988, Congress amended TSCA by adding Title III—Indoor Radon Abatement (15 U.S.C. 2661 *et seq.*, P.L. 100-551). The basic purpose of Title III is to provide financial and technical assistance to the states that choose to support radon monitoring and control; neither monitoring nor abatement of radon is required by the act.

Title III required EPA to update its pamphlet "A Citizen's Guide to Radon," to develop model construction standards and techniques for controlling radon levels within new buildings, and to provide technical assistance to states. EPA is to provide technical assistance by establishing an information clearinghouse; publishing public information materials; establishing a national database of radon levels detected, organized by state; providing information to professional organizations representing private firms involved in building design and construction; submitting to Congress a plan for providing financial and technical assistance to states; operating cooperative projects with states; conducting research to develop, test, and evaluate radon measurement methods and protocols; developing and demonstrating new methods of radon measurement and mitigation, including methods that are suitable for use in nonresidential child care facilities; operating a voluntary program to rate radon measurement and mitigation devices and methods and the effectiveness of private firms and individuals offering radon-related services; and designing and implementing training seminars. The proficiency rating program and certification for training programs collect fees for service, and therefore are meant to be self-supporting, but Congress authorized $1.5 million to be appropriated to establish these programs. Congress authorized $3 million to be appropriated for each of three years beginning in 1989 for the other provisions of Sections 303, 304, and 305.

A matching grant program was established for the purpose of assisting states in developing and implementing programs for radon assessment and mitigation. For this program, $30 million was authorized to be appropriated over three years, with funds targeted to states or projects that made efforts to ensure adoption of EPA's model construction standards and techniques for new buildings; gave preference to low-income persons; or addressed serious and extensive radon contamination problems or had the potential to reduce risk or to develop innovative assessment techniques, mitigation measures, or management approaches.

Other sections of Title III require EPA to conduct a study to determine the extent of radon contamination in schools; identify and list areas of the United States with a high probability of having high levels of indoor radon; make grants or cooperative agreements to establish and operate at least three regional radon training centers; and provide guidance to federal agencies on radon measurement, risk assessment, and remedial measures.

All authorizations for appropriations specific to this title expired September 30, 1991, although appropriations have continued.

TITLE IV (LEAD EXPOSURE REDUCTION)

The 102nd Congress added Title IV to TSCA when it enacted the Residential Lead-Based Paint Hazard Reduction Act of 1992 as Title X in the Housing and Community Development Act of 1992 (P.L. 102-550). TSCA Title IV aims to accelerate federal efforts to reduce risks to young children who daily are exposed to lead-based paint in their homes. In addition, it was intended to stimulate development of lead inspection and hazard abatement services in the private sector, while ensuring that the services provided and any products employed are reliable and effective in reducing risk. To these ends, Title IV directs EPA to

- promulgate definitions of lead-contaminated dust, lead-contaminated soil, and lead-based paint hazards;

- ensure that people engaged in detection and control of lead hazards are properly trained and that contractors are certified;
- publish requirements for the accreditation of training programs for workers;
- develop criteria to evaluate the effectiveness of commercial products used to detect or reduce risks associated with lead-based paint;
- establish protocols, criteria, and minimum performance standards for laboratory analysis of lead in paint films, soil, and dust;
- establish a program to certify laboratories as qualified to test substances for lead content; and
- publish and distribute to the public a list of certified or accredited environmental sampling laboratories.

Title IV explicitly applies these requirements to federal facilities and activities that may create a lead hazard.

In addition, Congress directed EPA to conduct a study of lead hazards due to renovation and remodeling activities that may incidentally disturb lead-based paint. EPA is required to promulgate guidelines for the renovation and remodeling of buildings or other structures when these activities might create a hazard.

Title IV directs EPA to establish a clearinghouse and hotline to distribute information about the hazards of lead-based paint, how to avoid exposure and reduce risk, and new technologies for removing or immobilizing lead-based paint. In addition, Congress mandated development of a lead hazard information pamphlet; public education and outreach activities for health professionals, the general public, homeowners, landlords, tenants, consumers of home improvement products, the residential real estate industry, and the home renovation industry; and information to be distributed by retailers of home improvement products to provide consumers with practical information related to the hazards of renovation where lead-based paint may be present.

Title IV authorizes states to propose programs to train and certify inspectors and contractors engaged in the detection or control of lead-based paint hazards. States also may develop the required informational pamphlets. TSCA requires EPA to promulgate a model state program that may be adopted by any state. Congress gave EPA the authority to approve or disapprove authorization for state proposals and to provide grants for states to develop and implement authorized programs. A federal program must be established, administered, and enforced by EPA in each state without an authorized program.

The Department of Health and Human Services also has responsibilities under Title IV of TSCA. It mandates a study by the Centers for Disease Prevention and Control (CDC) and the National Institute for Environmental Health Sciences to determine the sources of lead exposure to children who have elevated lead levels in their bodies. The National Institute for Occupational Safety and Health is directed to study ways of reducing occupational exposure to lead during abatement activities.

The act established a rule-making docket to ensure the availability to the general public of all documents submitted to agencies that are relevant to regulatory decisions pursuant to this legislation. The docket is required to include the drafts of all proposed rules submitted by EPA to the President's Office of Management and Budget (OMB), written comments on the drafts, and written responses to comments. In addition, the agency must provide an explanation for any major change to a proposed rule that appears in the final rule, and such changes may not be made based on information not filed in the docket. Dockets are required to be established in each EPA regional office.

Congress authorized to be appropriated "such sums as may be necessary" for TSCA Title IV.

In addition to amending TSCA, Title X of the Housing and Community Development Act of 1992 authorized grants to states for risk assessments and lead-based paint removal and immobilization in private housing for low-income residents; establishing state training, certification, or accreditation programs for inspectors and abatement contractors; and research at the Department of Housing and Urban Development (HUD). Authorization for appropriations for these grants expired September 30, 1994, but appropriations have continued. Title X directed HUD to establish guidelines for federally supported work involving risk assessments, inspections, interim controls, and abatement of lead-based paint hazards. In addition, the National Institute for Occupational Safety and Health (NIOSH) was provided $10 million for training people who remove or immobilize paint.

TITLE V (REDUCING RISKS IN SCHOOLS)

At the end of 2007, the 110th Congress added a fifth title to TSCA, subtitled Healthy High-Performance Schools. Enacted as Title IV, Subtitle E (§461) of P.L. 110-140, the Energy Independence and Security Act of 2007, TSCA Title V authorizes EPA to establish a state grant program to provide technical assistance for EPA programs to schools and develop and implement state school environmental health programs. State programs must include standards for school building design, construction, and renovation, and identify ongoing school building environmental problems and recommended solutions. Environmental problems specifically mentioned in the law include "contaminants, hazardous substances, and pollutant emissions." EPA's authority to provide grants expires five years after the date of enactment.

Title V requires the EPA Administrator, in consultation with the Secretary of Education and the Secretary of Health and Human Services, to issue voluntary guidelines within 18 months of Title V enactment for selecting sites for schools (presumably new schools). The guidelines are to account for the "special vulnerability of children to hazardous substances or pollution exposures in any case in which the potential for contamination at a potential school site exists," modes of transportation available to students and staff, efficient use of energy, and potential use of a school at the site as an emergency shelter.

Title V also requires the EPA Administrator, in consultation with the Secretary of Education and the Secretary of Health and Human Services, to issue voluntary guidelines within two years of enactment for developing and implementing state environmental health programs for schools. These guidelines must take into account the findings of federal initiatives established under "relevant federal law with respect to school facilities," including initiatives related to water and energy conservation authorized by Sections 431 through 441, and work related to high-performance green buildings authorized by Section 492 of P.L. 110-140. In particular, the guidelines must take into account "environmental problems, contaminants, hazardous substances, and pollutant emissions"; natural day lighting; ventilation; heating and cooling; moisture control and mold; maintenance, cleaning, and pest control; acoustics; and "other issues relating to the health, comfort, productivity, and performance of occupants of the school facilities." In addition, Title V requires that the guidelines provide "technical assistance on siting, design, management, and operation of school facilities"; collaborate with children's environmental health centers in school environmental investigations"; assist states and the public to better understand and improve the environmental health of children; and take into account "the special vulnerability of children in low-income and minority communities to exposures from contaminants, hazardous substances, and pollutant emissions."

Several provisions in Title V refer to entities established under other sections of the Energy Independence and Security Act of 2007 (P.L. 110-140). For example, Title V contains directives for the Federal Director of the Office of Federal High-Performance Green Buildings in the General Services Administration, which was created by Section 436(a). In addition, there is reference to the national high-performance green building clearinghouse established in Section 423(1) "to carry out public outreach to inform individuals and entities of the information and services [related to high-performance green buildings] available governmentwide." Title V requires the Federal Director to ensure, "to the maximum extent practicable," that the public clearinghouse "receives and makes available information on the exposure of children to environmental hazards in school facilities." The EPA Administrator is directed to prepare an annual report to Congress on activities carried out under Title V authority, and this report also must be made available to the public through the clearinghouse.

For the purposes of carrying out the provisions of Title V, Congress authorized appropriations of $1.5 million annually through 2013.

TITLE VI (LIMITING FORMALDEHYDE EMISSIONS)

In July 2010, Congress enacted the Formaldehyde Standards for Composite Wood Products Act (P.L. 111-199), adding a new Title VI to TSCA. The new title mandates specific formaldehyde emission standards for hardwood plywood, medium-density fiberboard, and particleboard that is sold, supplied, offered for sale, or manufactured in the United States. The standards are phased in over two years from enactment and are based on the voluntary national formaldehyde emissions standards established by ASTM International (formerly known as the American Society for Testing and Materials), method ASTM E-1333-96 (2002).

The standards apply to plywood, particleboard, and medium-density fiberboard in the form of an unfinished panel or incorporated into a finished good. Certain products are excluded, including many forms of lumber and panels used for outdoor applications, such as structural plywood, prefabricated wood I-joists, most windows, antiques or other previously owned goods, and composite wood products used inside automobiles, trucks, rail cars, boats, and aircraft.

EPA is required to promulgate regulations ensuring compliance with the emission standards and must include provisions relating to labeling, chain of custody requirements, sell-through provisions; ultra low-emitting formaldehyde resins, finished goods, third-party testing and certification; auditing and reporting of third-party certifiers; recordkeeping; enforcement, laminated products; and exceptions for products and components containing "de minimis amounts" of composite wood products. The new law prohibits stockpiling of products manufactured before the effective date of the act for sale after that date. Also prohibited is any requirement for labeling products manufactured prior to the "designated date of manufacture."

P.L. 111-199 requires an annual report to Congress on the status of implementation and the extent to which relevant industries have achieved compliance. Finally, the act directs the Secretary of Housing and Urban Development to update regulations concerning formaldehyde emissions from composite wood in manufactured homes (24 Code of Federal Regulations 3280.308) to ensure that the standards established by TSCA Title VI are implemented.

Table 2. Major U.S. Code Sections, Toxic Substances Control Act (codified as 15 U.S.C. 2601-2692)

15 U.S.C.	Section Title	Toxic Substances Control Act (as amended)
Subchapter I	Control of Toxic Substances	
2601	Findings, policy and intent	§2
2602	Definitions	§3
2603	Testing of chemical substances and mixtures	§4
2604	Manufacturing and processing notices	§5
2605	Regulation of hazardous chemical substances and mixtures	§6
2606	Imminent hazards	§7
2607	Reporting and retention of information	§8
2608	Relationship to other federal laws	§9
2609	Research, development, collection, dissemination, and utilization of data	§10
2610	Inspections and subpoenas	§11
2611	Exports	§12
2612	Entry into customs territory of the United States	§13
2613	Disclosure of data	§14
2614	Prohibited acts	§15
2615	Penalties	§16
2616	Specific enforcement and seizure	§17
2617	Preemption	§18
2618	Judicial	§19
2619	Citizens' civil actions	§20
2620	Citizens' petitions	§21

Table 2. (Continued)

15 U.S.C.	Section Title	Toxic Substances Control Act (as amended)
2621	National defense waiver	§22
2622	Employee protection	§23
2623	Employment effects	§24
2624	Studies	§25
2625	Administration	§26
2627	Development and evaluation of test methods	§27
2628	Authorization of appropriations	§28
2629	Annual report	§29
Subchapter II	Asbestos Hazard Emergency Response	
2641	Congressional findings and purpose	§201
2642	Definitions	§202
2643	EPA regulations	§203
2644	Requirements if EPA fails to promulgate regulations	§204
2645	Submission to state Governor	§205
2646	Contractor and laboratory accreditation	§206
2647	Enforcement	§207
2648	Emergency authority	§208
2649	State and federal law	§209
2650	Asbestos contractors and local educational agencies	§210
2651	Public protection	§211
2652	Asbestos ombudsman	§212
2653	EPA study of asbestos-containing material in public buildings	§213
2654	Transition rules	§214
2655	Worker protection	§215
Subchapter III	Indoor Radon Abatement	
2661	National goal	§301
2662	Definitions	§302
2663	EPA's citizen guide	§303
2664	Model construction standards and techniques	§304
2665	Technical assistance to states for radon programs	§305
2666	Grant Assistance to states for radon programs	§306
2667	Radon in schools	§307
2668	Regional radon training centers	§308
2669	Study of radon in federal buildings	§309
2670	Regulations	§310
2671	Additional authorizations	§311
Subchapter IV	Lead Exposure Reduction	

15 U.S.C.	Section Title	Toxic Substances Control Act (as amended)
2681	Definitions	§401
2682	Lead-based paint activities training and certification	§402
2683	Identification of dangerous levels of lead	§403
2684	Authorized state programs	§404
2685	Lead abatement and measurement	§405
2686	Lead hazard information pamphlet	§406
2687	Regulations	§407
2688	Control of lead-based paint at federal facilities	§408
2689	Prohibited acts	§409
2690	Relationship to other federal law	§410
2691	General provisions relating to administrative proceedings	§411
2692	Authorization of appropriations	§412
Subchapter V	Healthy High-Performance Schools	
2695	Grants for healthy school environments	§501
2695a	Model guidelines for siting of school facilities	§502
2695b	Public outreach	§503
2695c	Environmental health program	§504
2695d	Authorization of appropriations	§505
Subchapter VI	Formaldehyde Standards for Composite Wood Products	
2697	Formaldehyde Standards	§601

Note: This table shows only the major code sections. For more detail and to determine when a section was added, the reader should consult the official printed version of the U.S. Code.

Selected References

Bergeson, Lynn L. TSCA: Toxic Substances Control Act. Basic Practice Series. Chicago, IL: Section of Environment, Energy, and Resources, *American Bar Association*, 2000. 151 p.

Bergeson, Lynn L., Lisa M. Campbell, and Lisa Rothenberg. TSCA and the Future of Chemical Regulation, *EPA Administrative Law Reporter*, 2000, v. 15, n. 4, 23 p.

CRS Report RS21688, Lead-Based Paint Poisoning Prevention: Summary of Federal Mandates and Financial Assistance for Reducing Hazards in Housing, by Linda-Jo Schierow.

In: Environmental Protection Agency ... ISBN: 978-1-62808-244-9
Editor: Robin Laurits

Chapter 3

ESSENTIAL PRINCIPLES FOR REFORM OF CHEMICALS MANAGEMENT LEGISLATION*

U.S. Environmental Protection Agency

The U.S. Environmental Protection Agency (EPA) is committed to working with the Congress, members of the public, the environmental community, and the chemical industry to reauthorize the Toxic Substances Control Act (TSCA). The Administration believes it is important to work together to quickly modernize and strengthen the tools available in TSCA to increase confidence that chemicals used in commerce, which are vital to our Nation's economy, are safe and do not endanger the public health and welfare of consumers, workers, and especially sensitive sub-populations such as children, or the environment.

The following Essential Principles for Reform of Chemicals Management Legislation (Principles) are provided to help inform efforts underway in this Congress to reauthorize and significantly strengthen the effectiveness of TSCA. These Principles present Administration goals for updated legislation that will give EPA the mechanisms and authorities to expeditiously target chemicals of concern and promptly assess and regulate new and existing chemicals.

* This document was released September 29, 2009.

Principle No. 1: Chemicals Should Be Reviewed Against Safety Standards That Are Based on Sound Science and Reflect Risk-based Criteria Protective of Human Health and the Environment.

EPA should have clear authority to establish safety standards that are based on scientific risk assessments. Sound science should be the basis for the assessment of chemical risks, while recognizing the need to assess and manage risk in the face of uncertainty.

Principle No. 2: Manufacturers Should Provide EPA With the Necessary Information to Conclude That New and Existing Chemicals Are Safe and Do Not Endanger Public Health or the Environment.

Manufacturers should be required to provide sufficient hazard, exposure, and use data for a chemical to support a determination by the Agency that the chemical meets the safety standard. Exposure and hazard assessments from manufacturers should be required to include a thorough review of the chemical's risks to sensitive subpopulations.

Where manufacturers do not submit sufficient information, EPA should have the necessary authority and tools, such as data call in, to quickly and efficiently require testing or obtain other information from manufacturers that is relevant to determining the safety of chemicals. EPA should also be provided the necessary authority to efficiently follow up on chemicals which have been previously assessed (e.g., requiring additional data or testing, or taking action to reduce risk) if there is a change which may affect safety, such as increased production volume, new uses or new information on potential hazards or exposures. EPA's authority to require submission of use and exposure information should extend to downstream processors and users of chemicals.

Principle No. 3: Risk Management Decisions Should Take into Account Sensitive Subpopulations, Cost, Availability of Substitutes and Other Relevant Considerations.

EPA should have clear authority to take risk management actions when chemicals do not meet the safety standard, with flexibility to take into account a range of considerations, including children's health, economic costs, social benefits, and equity concerns.

Principle No. 4: Manufacturers and EPA Should Assess and Act on Priority Chemicals, Both Existing and New, in a Timely Manner.

EPA should have authority to set priorities for conducting safety reviews on existing chemicals based on relevant risk and exposure considerations. Clear, enforceable and practicable deadlines applicable to the Agency and industry should be set for completion of chemical reviews, in particular those that might impact sensitive sub-populations.

Principle No. 5: Green Chemistry Should Be Encouraged and Provisions Assuring Transparency and Public Access to Information Should Be Strengthened.

The design of safer and more sustainable chemicals, processes, and products should be encouraged and supported through research, education, recognition, and other means. The goal of these efforts should be to increase the design, manufacture, and use of lower risk, more energy efficient and sustainable chemical products and processes.

TSCA reform should include stricter requirements for a manufacturer's claim of Confidential Business Information (CBI). Manufacturers should be required to substantiate their claims of confidentiality. Data relevant to health and safety should not be claimed or otherwise treated as CBI. EPA should be able to negotiate with other governments (local, state, and foreign) on appropriate sharing of CBI with the necessary protections, when necessary to protect public health and safety.

Principle No. 6: EPA Should Be Given a Sustained Source of Funding for Implementation.

Implementation of the law should be adequately and consistently funded, in order to meet the goal of assuring the safety of chemicals, and to maintain public confidence that EPA is meeting that goal. To that end, manufacturers of chemicals should support the costs of Agency implementation, including the review of information provided by manufacturers.

In: Environmental Protection Agency ...
Editor: Robin Laurits
ISBN: 978-1-62808-244-9

Chapter 4

EXISTING CHEMICALS PROGRAM: STRATEGY*

U.S. Environmental Protection Agency

OVERVIEW

The TSCA inventory of chemicals in commerce now exceeds 84,000 chemicals. Periodic TSCA chemical data reporting indicates that there are approximately 7,000 chemicals currently produced at volumes of 25,000 pounds or greater. Under TSCA, EPA is charged with the responsibility of assessing the safety of these commercial chemicals and to act upon those chemicals if there are significant risks to human health or the environment. A sustained and predictable approach is needed to effectively carry out this responsibility.

EPA believes that this significant and long-term challenge can best be met via legislative reform of TSCA to improve EPA's chemical management authorities[1]. Until reform is achieved, however, EPA's responsibility to create a sustained and effective existing chemicals program must be carried out under current authorities.

This strategy summarizes EPA's approach in 2012 and beyond, pending legislative reform.

---

* This is an edited, reformatted and augmented version of the U.S. Environmental Protection Agency publication, dated February 2012.

Given the vast number of chemicals; the high cost to EPA of performing comprehensive risk assessments, and, if appropriate, risk management; and the Agency's responsibility to protect human health and the environment, EPA has developed the following multi-pronged approach for the Agency's existing chemicals management program:

4. Risk assessment and risk reduction
5. Data collection and screening
6. Public access to chemical data and information

In summary, the Agency will perform risk assessments and, if appropriate, risk management for those chemicals with well-characterized hazard concerns and which present the possibility of significant exposure. These are likely to be a relatively small number of chemicals, compared to the size of the universe of commercial chemicals. While risk assessments are being conducted for this small group of chemicals, EPA will be developing an approach to screen the thousands of other compounds to determine which ones warrant further attention, which could include comprehensive risk assessments, or additional data development, addressing either hazard or exposure. Many chemicals will likely be judged as being of lower concern. Finally, EPA will work toward making chemical information available. In particular, the Agency will work to ensure that hazard and exposure data are available to the public in a manner that is most useful to those in the public who will be using the information. Taking this approach to address multiple aspects of the chemicals management challenge simultaneously should allow the Agency to be more comprehensive in its efforts, despite the large number of high production chemicals.

I. Risk Assessment and Risk Reduction

At present, there are thousands of chemicals that have yet to be screened and assessed. There are, however, a number of chemicals for which there are well-characterized hazard concerns and which present the possibility of significant exposure. It is important that EPA identify and assess these chemicals, even as the Agency works to collect data and to screen the many other chemicals in commerce for future assessment.

In 2011, EPA sought stakeholder input on the criteria for identifying such chemicals for the risk assessment/risk reduction component of EPA's work plan. The Agency heard general agreement with the proposal to identify chemicals for assessment with well-characterized concerns for human health or environmental toxicity, which are persistent and bioaccumulative, are used in consumer or children's products, have dispersive uses, or have been detected in human or environmental biomonitoring. EPA also received stakeholder input on the use of a number of well-known, well-documented sources proposed to inform that process.[2]

EPA has used these criteria to identify a work plan of 83 chemicals for review. From this work plan list, EPA has identified an initial set of chemicals for which EPA has begun risk assessments in 2012. The Agency will complete some and initiate additional new assessments each succeeding year. The chemical assessments may also include evaluation of alternatives. If an assessment indicates significant risk, EPA will evaluate and pursue appropriate risk reduction actions. If an assessment indicates no significant risk, EPA will conclude its current work on that chemical. Over time, additional chemicals will be added to the work plan as more data are developed and more chemicals screened.

In 2012, the Agency will also continue to carry out work identified for chemicals in Action Plans, under other statutory requirements such as those for formaldehyde in pressed wood products, and other priority issues as they arise.

DATA COLLECTION AND SCREENING

For the thousands of chemicals for which toxicity or exposure data are less adequate to inform assessments, EPA must acquire and review data to support further screening and determine whether these chemicals are candidates for risk assessment and reduction, or whether they are of lower priority. Although, given current authorities and funding, it will take a number of years to work through the initial work plan of chemicals for assessment, EPA must begin now to build the pipeline of future assessment chemicals, as well as to identify chemicals of lower concern.

In 2011, EPA completed final amendments to the Chemical Data Reporting rule for the reporting of production and use information under that rule which will occur in 2012. The 2012 reporting will provide the first update on chemicals in commerce above the 25,000 pound threshold since 2006, and will provide use and exposure information on an expanded subset of chemicals. As information is being reported in 2012, EPA intends to engage with stakeholders to discuss how the CDR data might be used to aid the screening of chemicals to refine data needs and identify candidate chemicals for risk assessment and reduction. EPA plans to begin releasing data from the CDR within months of the end of the reporting period and to conduct and release further analysis of the data by early 2013.

Also in 2011, the Agency finalized several test rules for high production volume chemicals and proposed the fourth and final test rule in that series. In the proposed test rule, EPA solicited comment on future testing approaches by asking, for example: whether production volume should continue to drive testing requirements or whether other exposure or hazard factors should be included; whether the Screening Information Data Set remains the best testing approach; and how computational toxicology should be incorporated into the testing program. In 2012, EPA plans to engage stakeholders in discussion about the future approach to the testing program, and will continue to work on how best to utilize a range of data sources, with particular focus on data collected by other national governments and by states, also engaging stakeholders in these issues.

The Agency's approach to screening chemicals has involved a resource-intensive process to review in vivo data, as well as making estimates of exposure, to determine which chemicals warrant further review and attention. After screening, the Agency may decide a more comprehensive risk assessment is appropriate. In 2012 and 2013, the Agency will explore the use of computational techniques as a potentially faster means to screen thousands of chemicals far more quickly and efficiently.

While the pace of this work will depend on funding levels, the work in 2012 will put the Agency in a position to begin work in 2013 to identify chemicals for additional data collection and analysis and to begin the creation of a pipeline of candidate chemicals for future risk assessment and reduction.

Public Access to Chemical Data and Information

An important underpinning of a credible chemicals program is EPA's ability to make health and safety information available to the public, to the extent allowed by law. Since 2009, EPA has undertaken a number of specific efforts to increase the public's ability to access EPA's chemical information, e.g., the new Chemical Data Access Tool, free on-line access to the complete (non-confidential) TSCA Inventory, and including critical chemical information in Data.Gov. As part of a sustainable and predictable chemicals program, the Agency will continue to improve the accessibility and usability of its chemical data, including hazard and exposure information. This information will support the work of the Agency as well as informing other decision makers in the public and private sector.

Since 2009, EPA has made a priority the clearing of a large backlog of unchallenged CBI claims in health and safety studies, so that only justified CBI is withheld from the public. Industry has acknowledged that excessive claims of confidentiality have been made over the years of TSCA implementation, as EPA failed to require justification for such claims. In addition, outdated program systems have limited the ready accessibility of data to both Agency staff and the public. The Agency has also begun to digitize the TSCA data holdings and to create search tools that improve access.

In 2012, EPA will continue to work to address unsupported claims of confidentiality for chemicals in commerce, and to digitize and make public additional health and safety data from the TSCA data holdings. The Agency will seek stakeholder input on tools and approaches for improving the accessibility and usefulness of these data for private and public sector decision-making.

End Notes

[1] http://www.epa.gov/oppt/existingchemicals/pubs/principles.html
[2] http://www.epa.gov/oppt/existingchemicals/pubs/chempridiscguide.html

In: Environmental Protection Agency ... ISBN: 978-1-62808-244-9
Editor: Robin Laurits

Chapter 5

TSCA WORK PLAN CHEMICALS: METHODS DOCUMENT*

U.S. Environmental Protection Agency

BACKGROUND

In the Agency's August 2011 *Discussion Guide: Background and Discussion Questions for Identifying Priority Chemicals for Review and Assessment*, EPA described the two-step process the Agency intended to use to identify potential candidate chemicals for near-term review and assessment under the Toxic Substances Control Act (TSCA). The Agency intends to use these TSCA Work Plan Chemicals to help focus and direct the activities of the Existing Chemicals Program in the Office of Pollution Prevention and Toxics (OPPT). *EPA invited public comment* through an *online discussion forum* conducted from August 18 through September 21, 2011, as well as through a webinar and stakeholder meeting held on September 7, 2011. The meeting summaries and public comments are available for review in the docket for this activity, EPA-HQ-OPPT-2011-0516, which can be accessed online at http://www.regulations.gov.

As described in the *Discussion Guide*, EPA notes that identification of a chemical as a TSCA Work Plan Chemical does not itself constitute a finding by the Agency that the chemical presents a risk to human health or the

* This is an edited, reformatted and augmented version of the U.S. Environmental Protection Agency, dated February 2012.

environment. Such a determination would be the result of a risk assessment. Rather, identification of a chemical as a TSCA Work Plan Chemical indicates only that the Agency intends to consider it for further review. The Agency believes that identifying these chemicals early in the review process would afford all interested parties the opportunity to bring additional relevant information on those chemicals to the Agency's attention in order to further inform the review. In order to take risk management actions on a chemical substance under various sections of TSCA, the Agency would have to make the appropriate findings required by the specific provisions of the statute.

Identification of some chemicals as TSCA Work Plan Chemicals (Work Plan) does not mean that EPA would not consider other chemicals for risk assessment and potential risk management action under TSCA and other statutes. EPA will consider other chemicals if warranted by available information. In addition, EPA may subsequently identify other candidates for review in addition to this initial group, and may adapt the factors and data sources used in this process based on the experience acquired during this initial phase. Further, while the chemicals identified through this process as TSCA Work Plan Chemicals will likely be well-characterized for hazard and have information indicating exposure potential, some will have more limited data and EPA will continue to use its TSCA information collection, testing, and subpoena authorities, including sections 4, 8, and 11(c) of TSCA, to develop needed information on additional chemicals that currently have less robust hazard or exposure databases.

TWO-STEP PROCESS

As described in the *Discussion Guide*, EPA's two-step prioritization process was intended to select an initial group of candidate chemicals for review by using a specific set of data sources to identify chemicals meeting one or more of the following factors:

- Chemicals identified as potentially of concern for children's health (e.g., chemicals with reproductive or developmental effects).
- Chemicals identified as persistent, bioaccumulative, and toxic (PBT).
- Chemicals identified as probable or known carcinogens.
- Chemicals used in children's products.
- Chemicals used in consumer products.
- Chemicals detected in biomonitoring programs.

EPA indicated the candidate chemicals from Step 1 would then be screened in Step 2 using information from additional exposure and hazard data sources to further analyze the chemicals and select specific chemicals for further assessment, including possible risk assessment and risk management action.

Based on comments received through the discussion forum, the webinar, and the stakeholder meeting, EPA made some adjustments both to the Step 1 factors and to the data sources utilized in both Step 1 and Step 2. With regard to the factors considered in Step 1, EPA added neurotoxicity to the initial Step 1 selection criteria because of comments noting the importance of neurotoxic effects to children's health. The Agency further added respiratory sensitization to the human health factors it would consider in Step 2, based on public comments suggesting this endpoint as identifying possible contributors to childhood asthma. Several commenters also encouraged EPA to use environmental toxicity as a prioritization factor to populate the Step 1 group of candidate chemicals. While environmental toxicity is not being used as a Step 1 prioritization factor on its own, EPA notes that many of the PBT chemicals are classed as toxic on the basis of environmental toxicity data. The Agency has also specifically factored environmental toxicity into the Step 2 analysis.

Following public comment, EPA also adjusted the proposed data sources identified in the *Discussion Guide,* particularly for Step 2, to encompass additional sources suggested by commenters, including the European Chemical Substance Information System (ESIS) and the Organization for Economic Cooperation (OECD) eChem Portal (which includes U.S. databases). EPA also eliminated certain data sources, including NHATS, NHEXAS, and TEAM, on the basis of their age. Given the difficulty of comprehensively identifying chemicals in consumer products, particularly because the 2006 Inventory Update Reporting (IUR) system made no distinction between commercial and consumer products, EPA narrowed the focus of the Step 1 prioritization factor to chemicals identified as being in children's products either through IUR reporting or through the process used by Washington State to generate its list of children's product chemicals. EPA notes, however, that chemicals identified through the application of the prioritization factors in Step 1 were further scrutinized in Step 2 against additional databases including the Hazardous Substance Data Bank (HSDB) and the Household Product Database, among others, to identify potential consumer uses.

DERIVATION OF THE STEP 1 POTENTIAL CANDIDATE CHEMICALS

To generate the Step 1 chemicals meeting the Agency's prioritization factor criteria as potential candidates for review and assessment, the following sources were used:

- ***Carcinogenicity:***
    - IRIS: 1986 Class A, B1; 1996 Known or Probable; 1999 or 2005 Carcinogenic
    - IARC Carcinogens, Group 1, 2A
    - NTP Known Carcinogens
- ***PBT:***
    - TRI PBT Rule
    - Great Lakes Binational PBT
    - Canadian P, B, and T (all three criteria met)
    - LRTAP POPS
    - Stockholm POPs
- ***Children's Health:***
    - IRIS: Repro/Dev (RfD or RfC for repro or dev)
    - NTP CERHR: Infants Any Effect or Pregnant Women Any Effect
    - Cal Prop 65 Reproductive
- ***Neurotoxicity: IRIS***
- ***Children's Product Use:***
    - Reported in products intended for use by children in 2006 IUR
    - Washington State Children's List
- ***Biomonitoring*** (both human and environmental indicative of potential human exposure):
    - NHANES
    - Drinking Water Contaminants
    - Fish Tissue Studies

These sources produced a combined total of 1,235 chemicals, each of which matched at least one criterion. The resulting chemicals were then screened both for quality control to eliminate duplicate listings (an artifact of differences in the way the various data sources defined and reported chemicals), and to exclude chemicals that would not be appropriate for designation as candidates for near-term review and action under TSCA, either because they did not meet the intent of the prioritization criteria, they were not subject to action under TSCA, or they were already the subject of TSCA action.

Chemicals were excluded from identification as potential candidates for any of the following reasons:

- **Pesticides:** Pesticides are excluded from regulation under TSCA because they are regulated under the Federal Insecticide, Fungicide, and Rodenticide Act (FIFRA).
- **Drugs, hormones, and pharmacological chemicals:** Drugs are excluded from regulation under TSCA because they are regulated under the Federal Food, Drug, and Cosmetic Act (FFDCA). Hormones and pharmacological chemicals can be found in the environment when they are excreted or disposed of, but may not be amenable to management under TSCA.
- **Certain radioactive materials:** Radioactive chemicals are generally excluded from regulation under TSCA as source materials, special nuclear materials, or byproduct materials as defined in the Atomic Energy Act and subsequent regulations.
- **Complex process streams, byproducts not commercially produced:** Chemicals that are the reaction products of vague constituents, byproducts of complex streams, or complex mixtures are generally not readily definable in terms of their chemical identity and may vary considerably in both their composition and hazard from batch to batch, making them difficult to score consistently in this type of screening exercise. They were accordingly excluded.
- **Polymers:** Polymers typically have physical and chemical characteristics (high molecular weight, low absorbance, and low reactivity) that do not generally present significant health hazards. Some polymers that meet certain established criteria (49 FR 46066, November 21, 1984) have been specifically exempted from TSCA review under the new chemicals program because they "do not present an unreasonable risk of injury to human health or the environment." Polymers were therefore excluded from the Work Plan.
- **Gases, common naturally occurring chemicals, combustion products:** Chemicals that exist in gaseous form at normal temperatures, predominantly occur naturally in the environment, or are produced predominantly by combustion are generally not amenable to control or management under TSCA.

- **Common oils or fats, simple plant extracts:** Chemicals in these categories are generally not anticipated to be sufficiently toxic to give rise to concerns that would make them priorities.
- **Explosive, pyrophoric, or extremely reactive or corrosive chemicals:** Chemicals that explode, burn on contact with air or water, react quickly with other chemicals, or are extremely corrosive are unlikely to present opportunities for human or environmental exposures because their high physical hazard properties make them subject to stringent handling requirements intended to guard against accidental exposures or releases.
- **Metals principally identified as toxic to the environment:** Many metals – copper, for example – are generally toxic to the environment, but do not present health issues to humans under typical conditions of use. Those metals and related compounds were excluded from the Work Plan, while metals with specific human health concerns were retained.
- **Chemicals already the subject of Action Plans or significant regulation under TSCA:** *Polychlorinated biphenyls (PCBs)* were excluded from the Work Plan because they are already comprehensively regulated under TSCA, which bans their manufacture, processing, use and distribution in commerce. *Chemicals covered by Action Plans* or other currently ongoing regulatory activities under TSCA were also excluded because they had been recently reviewed and are already being addressed.

After these chemicals were excluded and the remaining metals and their related compounds were grouped together rather than being identified separately, 345 chemicals remained as potential candidates and entered into Step 2, which is described in the next section of this paper.

EXPLANATION OF STEP 2 PROCESS

The chemicals identified as potential candidates for review and assessment under TSCA based on the Step 1 prioritization factors were screened in Step 2. Chemicals were evaluated and received a score through the application of a numerical algorithm. This score was based on three characteristics: hazard, exposure, and potential for persistence and/or bioaccumulation. Using this system, chemicals were sorted into one of four bins. Chemicals able to be scored on all three characteristics were scored as High, Moderate, or Low based on their available information. Chemicals with High or Moderate hazard or persistence/bioaccumulation scores that could not be scored for exposure because of an absence of data, together with chemicals that could not be scored for hazard, were identified separately as potential candidates for information gathering.

This chemical candidate screening process is an interim evaluation only. It does not constitute a final Agency determination as to risk or as to whether sufficient data are available to characterize risk from specific chemicals on the TSCA Work Plan. Inclusion of a chemical on the Work Plan does not constitute any finding of risk under TSCA. This screening process is intended only to support initial decisions to determine the relative priority for further assessments and to identify potential data needs for individual chemicals or chemical groups.

Hazard Score

The Hazard Score encompasses both human health and environmental toxicity concerns. The specific hazard classification criteria are based on the *Alternatives Assessment Criteria for Hazard Evaluation* developed by EPA's Design for the Environment Program (DfE). The DfE criteria for classifying the toxicity of specific chemicals were developed from authoritative sources including the United Nation's Globally Harmonized System (GHS) for Chemical Classification and Labeling and other EPA programs. The data determining the score for each chemical were obtained through the data sources identified in Appendix A. The hazard data reviews on each chemical were not exhaustive and do not rise to the level of assessments. Chemicals were scored on the basis of readily available data, and no judgment was made concerning gaps in or completeness of the available data set for a given chemical.

The Hazard Score was determined based on 3 hazard levels, and each hazard level had a corresponding hazard rank (High-3, Moderate-2, and Low-1). The concentration ranges or characteristics that correspond with each hazard level are listed in Table 1 below.

Candidate chemicals from Step 1 received a hazard rank score for each of the toxicity endpoints that were applicable based on the data readily available for each chemical. *The highest hazard rank score a chemical received for any single human health or environmental toxicity endpoint became its Hazard Score.* If the review on a chemical produced a High hazard score for any endpoint other than acute mammalian toxicity or acute or chronic aquatic toxicity, data on other endpoints were not sought because they would not impact the existing High score.

Because the highest score from any individual endpoint was taken as the total Hazard Score, a chemical was ranked as either 3 (High), 2 (Moderate), or 1 (Low) for hazard.

For the toxicity endpoints Acute Mammalian Toxicity, Reproductive Toxicity, Developmental Toxicity, Neurotoxicity, and Chronic Toxicity a range of values for each Hazard Level was assigned. These values appear in the DfE *Alternatives Assessment Criteria*. In some cases DfE has 5 distinct hazard levels. For this analysis, the "Very High" and "High" levels from DfE were grouped together to represent High on this scale and DfE's "Low" and "Very Low" levels were combined to form the criteria for a Low rank.

Table 1. Criteria for Determining Hazard Score

	High	Moderate	Low	Hazard Score
Ranking	**3**	**2**	**1**	
Chemical X				**(Highest score from any toxicity category)**
Acute Mammalian Toxicity				
Oral LD50 (mg/kg)	≤ 50 - 300	>300 - 2000	>2000	
Dermal LD50 (mg/kg)	≤ 200 - 1000	>1000 - 2000	>2000	
Inhalation LC50 (gas/vapor) (mg/L)	≤ - 10 2	>10 - 20	>20	
Inhalation LC50 (mist/dust) (mg/L/day)	≤ 0 1.0 5	>1.0 - 5	>5	
Carcinogenicity	GHS 1A, 1B, GHS2	Limited animal	Negative or SAR	
Mutagenicity/Genotoxicity	GHS 1A, 1B, GHS 2	Positive *in vivo* or *in vitro*	Negative	
Reproductive Toxicity				
Oral (mg/kg/day)	<50	50-250	>250	
Dermal (mg/kg/day)	<100	100-500	>500	
Inhalation (gas/vapor) (mg/L/day)	<1	1-2.5	>2.5	
Inhalation (mist/dust) (mg/L/day)	<0.1	0.1-0.5	>0.5	
Developmental Toxicity				
Oral (mg/kg/day)	<50	50 -250	>250	
Dermal (mg/kg/day)	<100	100- 500	>500	
Inhalation (gas/vapor) (mg/L/day)	<1.0	1.0 - 2.5	>2.5	
Inhalation (mist/dust) (mg/L/day)	<0.1	0.1- 0.5	>0.5	
Neurotoxicity				
Oral (mg/kg-bw/day) 90- day (13 weeks)	< 10	10- 100	> 100	
40-50 days	< 20	20-200	> 200	
28-days (4 weeks)	< 30	30- 300	> 300	
Dermal (mg/kg-bw/day) 90-day (13 weeks)	< 20	20- 200	> 200	
40-50 days	< 40	40 - 400	> 400	
28-days (4 weeks)	< 60	60 -600	> 600	

Table 1. (Continued)

	High	Moderate	Low	Hazard Score
Ranking	**3**	**2**	**1**	
Chronic Toxicity				
Oral (mg/kg-bw/day) 90-day (13 weeks)	< 10	10-100	> 100	
40-50 days	< 20	20-200	> 200	
28-days (4 weeks)	< 30	30-300	> 300	
Dermal (mg/kg-bw/day) 90-day (13 weeks)	< 20	20-200	> 200	
40-50 days	< 40	40-400	> 400	
28-days (4 weeks)	< 60	60-600	> 600	
Respiratory Sensitization	GHS 1A and 1B Occurrence of respiratory sensitization; Evidence supporting potential for respiratory sensitization		No evidence to support potential for respiratory sensitization	
Acute Aquatic Toxicity (LC50 or EC50) (mg/L)	< 1.0-10	> 10 - 100	> 100	
Chronic Aquatic Toxicity (NOEC or LOEC) (mg/L)	< 0.1-1	> 1 - 10	> 10	
				Hazard Score

The hazard levels for Carcinogenicity were based on whether a chemical is a known, presumed, or suspected carcinogen (High); limited evidence of carcinogenicity (Moderate); or non- carcinogenetic (Low). Note that the High score for carcinogenicity in Step 2 is broader than the criteria used in the Step 1 for carcinogenicity. The Step 1 factor specified that a chemical be classified as a known or probable carcinogen, equivalent to the GHS 1A or 1B classification, in order to be included in the screening program expressly on the basis of carcinogenicity. For the purpose of further evaluating the Agency's potential concern for chemical hazard in Step 2 of this screening process, however, EPA included presumed, suspected, or likely human carcinogenicity classifications – the equivalent of GHS 2 – as also meriting a High hazard score.

The hazard levels for Mutagenicity/Genotoxicity were based on evidence that heritable mutations are known to or may occur in human germ cells, or mutagenicity demonstrated *in vivo* and *in vitro* (High); evidence of mutagenicity supported by *in vivo* or *in vitro* somatic cells of humans and animals (Moderate); or no evidence of chromosomal aberrations and gene mutations in reported studies (Low).

Respiratory Sensitization was based on GHS classifications of respiratory sensitizers. Hazard levels were based on whether there is occurrence of respiratory sensitization in humans or supporting evidence based on other tests, including the presence of structural alerts (High); or no evidence to support the potential for respiratory sensitization (Low). This endpoint was added to the prioritization template proposed in the August 2011 *Discussion Guide* following stakeholder comment that respiratory sensitization is particularly of interest to children's health issues based on the increasing trends of childhood asthma and other illnesses.

Environmental toxicity information was limited primarily to aquatic toxicity studies. If information about environmental toxicity was available, it was analyzed in conjunction with human toxicity information.

Chemicals that were scored as High for hazard only on the basis of acute mammalian toxicity were further considered on the basis of their classification for other human health endpoints. Where data on other health endpoints were available, the overall hazard score for the chemical was adjusted accordingly to reflect the highest remaining health endpoint. This was done because chemicals with high acute mammalian toxicity are generally already regulated on the basis of that toxicity and are subject to handling and use controls intended to protect workers and others potentially coming into contact with the chemical from harmful acute exposures. Scoring those chemicals on the basis of their other toxic effects was intended to acknowledge that protection against effects from acute exposures would not necessarily protect against effects from other exposures. If acute mammalian toxicity was the only available data endpoint for a chemical, the acute score remained as the overall hazard score for the chemical.

Chemicals that scored as High for hazard only on the basis of acute or chronic aquatic toxicity but that did not present human health concerns were grouped separately as being of potential concern for the environment.

If no hazard data were available on a chemical to provide a hazard score, the chemical was placed in a parallel prioritization category. These chemicals were classified as "Potential Candidates for Information Gathering. (See page 16.)" Creating a separate category ensured that chemicals with unknown toxicity would not be removed from further investigation because there was a lack of data.

Exposure Score

The Exposure Score was based on a combination of chemical use, general population and environmental exposure, and release information. The Use Type score included consideration of consumer product applications as well as industrial and commercial uses that could result in widespread exposures. The General Population and Environmental Exposure score encompassed measured data on the presence of a chemical in biota and environmental media. The Release score was based on EPA's Toxics Release Inventory (TRI) data for chemicals subject to TRI reporting. For non-TRI chemicals, the Release score was calculated using a method involving Inventory Update Reporting data (IUR, now called Chemical Data Reporting, or CDR), including production volume, number of sites, and type of use. Data used in the other two components of exposure scoring were obtained through the sources identified in Appendix B. The detailed description of how information from those sources was used to generate an exposure score appears in Appendix C.

The criteria for exposure potential in the Use Types category were based on a chemical's presence and characteristics of use in consumer, commercial, or industrial products as indicated in the data sources in Appendix B. Chemicals in consumer products judged widely used with a high potential for exposure received the highest rank. Chemicals that are present in consumer products but are more narrowly used and have lower likelihood of exposure were ranked as moderate. Chemicals that are not high or moderate but have commercial uses reported in IUR were ranked as low, acknowledging that such uses may present some potential for exposures not only to workers but also to the general population and the environment. Chemicals with no commercial use reported in IUR received a rank of zero. Further information on this approach and examples of ranking by use type are provided in Appendix C.

The data supporting ranking in the General Population and Environmental Exposure category came from the databases and peer-reviewed studies included in the list presented in Appendix B. The highest rank was based on presence in biota, because chemicals measured in humans, fish, animals, or plants demonstrate clear evidence of exposure; and on measured presence in indoor air, house dust, or drinking water, because presence in those specific media provides a strong indication of exposure potential. Presence in two or more environmental media indicates a reasonable potential for environmental exposure, which was the criteria for a moderate exposure ranking. Measured presence in one environmental medium provides some indication of potential environmental exposure, and was given a low ranking.

The Release Score was determined in one of two ways. If the chemical was reported under TRI, the TRI data were used to infer potential for environmental and general population exposure. The breakdowns for the high, moderate and low ranks were based on a distribution of pounds released for the chemicals reported by industry in the database.

If no TRI data existed, a release score was calculated on the basis of IUR data using production volume, number of sites, and use codes classified according to how likely the uses were to result in releases. The description of how these non-TRI release scores were derived, along with examples of how IUR use codes were associated by EPA with high, moderate, or low potentials for release, appears in Appendix C. While a chemical's production volume, use type, and number of manufacturing, processing, and industrial use sites do not provide exposure data, they can be used as an indicator of potential releases and resulting potential exposures.

All Exposure category scores were added up and then normalized on an overall High- Moderate-Low scale. To prevent the prioritization process from being biased unduly either toward or against data-rich chemicals, the normalization process differed depending on how many of the three categories – Use Type, General Population & Environmental Exposure, and Releases – had sufficient data to provide a score for the category.

Table 2. Exposure Score

				Score
I. Use Type				
Ranking	Criteria			**Use Score**
3	Consumer product widely used, high likelihood of exposure			
2	Consumer product narrow use, lower likelihood of exposure			
1	Commercial use, indicating some likelihood of exposure			
0	No reported commercial use, indicating little to no likelihood of general exposure from use			
II.	**General Population and Environmental Exposure**			
Ranking	Criteria			**+ General Population & Environmental Exposure**
3	Present in biota (human, fish, animal or plant biomonitoring), OR measured in drinking water, indoor air, house dust			
2	Not in biota, but reported present in 2 or more environmental media			
1	Reported present in 1 environmental medium			
III. Release Score: Use III. A or III. B, As Appropriate				
III. A.	**Release Score for TRI Chemicals***			**+ TRI Release Score**
Ranking	Criteria			
3	> 100,000 lbs/year			
2	5,000-100,000 lbs/year			
1	< 5,000 lbs/year			
	OR			**OR**
III. B.	**Release Score for Non-TRI Chemicals**			**+ Non-TRI Release Score**
The III.B. Release Score for Non-TRI Chemicals was generated by normalizing the sum of the subset rankings for Production Volume, Number of Sites, Industrial Processing and Use, and Commercial/Consumer Use differentiating between uses with high, moderate, and low potential for widespread releases, as shown below and described in detail in Appendix C:				
	Subset 1: IUR Production Volume		**PV**	
	Ranking	Criteria		
	3	≥1,000,000		
	2	≥ 500,000 -999,999 lbs/year		
	1	< 500,000 lbs/year		
	Subset 2: IUR Number of Manufacturing, Processing, and Use Sites		**+ Site #**	
	Ranking	Criteria		
	3	≥ 1, 000		
	2	100- 999		
				Score
	1	< 100		

	Subset 3: IUR Industrial Processing and Use (IPU)		**+ Use1**	
	Ranking	Criteria		
	3	High potential for release		
	2	Moderate potential for release		
	1	Low potential for release		
	Subset 4: IUR Commercial Use (C)		+Use2	
	Ranking	Criteria		
	3	High potential for release		
	2	Moderate potential for releases		
	1	Low potential for release		
	Subtotal Surrogate Score			
Total			=	**Exposure Score****

* TRI data included in the exposure calculation were limited to water, air, and non-contained land releases.

** Total Exposure Score is the sum of the individual scores for I, II, and III.A or III.B.

For chemicals with scores in all three categories, "9" was the highest possible score, and the normalization scoring structure was:

Total Exposure Score from Table 2	Overall Rank	Normalized Overall Exposure Score
8 - 9	High	3
5 - 7	Moderate	2
2 - 4	Low	1

For chemicals with scores in only two of the three categories, "6" was the highest possible score, and the normalization scoring structure was:

Total Exposure Score from Table 2	Overall Rank	Normalized Overall Exposure Score
5 - 6	High	3
3 - 4	Moderate	2
1-2	Low	1

In the absence of exposure data on chemicals sufficient to populate at least two of the exposure categories in Table 2 and produce a meaningful score, such chemicals receiving moderate or high hazard scores, or that also could not be scored for hazard because of an absence of hazard data, were placed in a parallel prioritization category. These chemicals were classified as "Potential Candidates for Information Gathering. (See page 16.)" EPA created this separate category to ensure that chemicals with unknown toxicity or with known potential human health or environmental toxicity implications would not be removed from further investigation simply because there was a lack of exposure information, an issue stakeholders identified during the webinar and discussion forum as being of concern.

Potential for Persistence/Bioaccumulation

Chemicals received a separate score to rank their potential for persistence and/or bioaccumulation. Persistent and bioaccumulative chemicals present special issues because organisms can remain exposed to them for a very long time and organisms higher up the food chain may be exposed to larger quantities of the chemicals through their food supply. EPA considers it particularly important that these chemicals not be removed from consideration for further investigation simply because they may lack either hazard or exposure information, or both.

Persistence scoring consisted of the evaluation of the potential half-life in air, water, soil, and sediment while considering the expected partitioning characteristics of the chemicals and all potential removal pathways based on standard physical-chemical properties and environmental fate parameters. Data sources listed in Appendix B were searched to locate studies on biotic and abiotic transformation (e.g., biodegradation, hydrolysis, photolysis) in order to estimate half-lives for the chemicals in the environment.

Table 3. Persistence/Bioaccumulation Potential

		Overall Persistence/ Bioaccumulation Score
I. Persistence		
Ranking	Criteria	Persistence
3	Half-life > 6 months	
2	Half-life ≥ 2 months	
1	Half-life < 2 months	
II. Bioaccumulative Potential		
Ranking	Criteria	+ Bioaccumulation
3	BCF or BAF > 5000	
2	BCF or BAF ≥ 1000	
1	< 1000	
Total		**Persistence/ Bioaccumulation Score**

Bioaccumulation scoring consisted of evaluation of bioaccumulation/ bioconcentration (measured or estimated BAF/BCF) data. When BAF data were not available, bioconcentration data (measured or estimated) were used to evaluate the potential for a chemical to bioaccumulate in organisms in the environment.

In the absence of test data establishing the chemical's measured persistence or bioaccumulation potential, EPA used EPI Suite™ version 4.10 to derive a ranking for the chemical. Specifically, BIOWIN, HYDROWIN, AOPWIN, BCF/BAF and Level III fugacity models were used to assess biodegradation, hydrolysis, atmospheric oxidation, bioaccumulation/ bioconcentration and environmental partitioning.

These criteria for judging persistence and bioaccumulation are the ones used in EPA's New Chemicals program. The separate scores for persistence and bioaccumulation were added together to produce a total score, which was normalized as follows:

Persistence/Bioaccumulation Score	Ranking	Normalized P/B Score
5 - 6	High	3
3 - 4	Moderate	2
2	Low	1

Categorizing Candidates for Inclusion as TSCA Work Plan Chemicals

After the candidate chemicals in Step 1 received normalized scores for Hazard, Exposure, and Persistence/Bioaccumulation, those scores were totaled to roughly group the chemicals receiving scores in all three categories into High, Moderate, and Low groupings as follows:

Normalized Total Score	Ranking
7 - 9	High
4 - 6	Moderate
1 - 3	Low

Appendix D identifies the 83 candidate chemicals from Step 1 that received scores on all three ranking factors and ranked High on the basis of their total score, including human health hazard concerns, and provides a brief summary of the information that produced that ranking. This table also includes chemicals that may not have presented human health concerns, but met all the criteria for identification as persistent, bioaccumulative, and environmentally toxic chemicals. These are the TSCA Work Plan Chemicals, from which the Agency intends to select chemicals for near-term review and assessment.

EPA notes that some chemicals identified as High through this scoring system may not necessarily be practical candidates for assessment under TSCA when other information is factored into the process. For example, the particular risks presented by certain chemicals may already be addressed by significant regulation under other statutes. One such example is quartz, which presents a hazard only in the context of silicosis from the inhalation of very fine crystalline dust particles, which could generally occur only during such occupational activities as sandblasting or stone cutting; these potential exposures are specifically controlled under regulations issued by the Occupational Safety and Health Administration (OSHA).

Potential Candidates for Information Gathering

Chemicals that could not be scored for hazard, or that were scored as moderate or high for either hazard or for persistence/bioaccumulation but could not be scored for exposure, have been grouped separately. These chemicals may be potential candidates for information-gathering activities focused on producing sufficient information to determine where they would rank in the prioritization process. EPA may consider a variety of such information-gathering activities, including both voluntary data submission and regulations issued under Sections 4 and 8 of TSCA.

Identifying Work Plan Chemicals for Risk Assessment in 2012 and Beyond

In identifying a smaller set of chemicals for work in any given year, EPA considers a number of factors, including:

- Whether the chemical was identified as a "High" ranking chemical.
- Whether the chemical reflects more than one of the factors identified in Step 1 (for example, chemicals that were identified as a potential

concern for children's health and also were persistent, bioaccumulative, and toxic) and whether each of the factors was covered by the set of chemicals. These factors included health and environmental hazards, children's health, use in consumer products and dispersive uses, persistence and bioaccumulation, and detection in biomonitoring and environmental monitoring.

- Whether certain chemicals, or groups of chemicals, would benefit from some preliminary work to assure that risk assessments are targeted and scoped appropriately, and therefore would best be addressed in an out year.
- Whether certain chemicals, or groups of chemicals, have previously been assessed and addressed by the Agency, so that risk assessment in later years may be more appropriate than in the earlier years of the work plan.
- Agency work load considerations, including scope and timing of work needed on specific chemicals, and existing commitments for assessment.

For 2012, EPA identified an initial group of seven chemicals, which can be found on the first page of the table in Appendix D. EPA will identify a group of chemicals each year for risk assessment, completing a number of risk assessments that year and initiating new assessments from the remaining chemicals on the work plan in the coming years. This spring, the Agency plans to identify specific chemicals for which it plans to conduct risk assessment in 2013 and 2014.

APPENDIX A. DATA SOURCES FOR HAZARD SCORING

Hazard Information (Data on all toxicological endpoints)	
Providers/ Data Source	**Description**
USEPA: IRIS	Integrated Risk Information System (IRIS): http://www.epa.gov/iris/index.html
USEPA: HPVIS	Hazard Characterizations prepared by EPA on chemicals in the High Production Volume Challenge Program (HPV): http://iaspub.epa.gov/oppthpv/hpv_hc_characterization.get_report?doctype=2 Risk-Based or Hazard-Based Prioritizations prepared by EPA under the Chemical Assessment and Management Program (ChAMP): http://iaspub.epa.gov/oppthpv/existchem_hpv_prioritizations.report

(Continued)

Providers/ Data Source	Description
Hazard Information (Data on all toxicological endpoints)	
USEPA: ISIS	The Integrated Scientific Information System (ISIS) is a chemical relational database application originally developed by Molecular Design Limited (MDL) Information Systems and utilized by the EPA New Chemicals program; the EPA version of this database contains confidential information.
United Nations World Health Organization: IARC	International Agency for Research on Cancer (IARC): http://monographs.iarc.fr/ENG/Classification/index.php
National Toxicology Program	NTP Report on Carcinogens: http://ntp.niehs.nih.gov/?objectid=03C9AF75-E1BF-FF40-DBA9EC0928DF8B15 NTP/CERHR Monographs on Potential Reproductive and Developmental Effects: http://ntp.niehs.nih.gov/?objectid=974B2C24-030F-D308-60E11D088F83FADB
Organization for Economic Cooperation and Development (OECD): eChem Portal	http://www.echemportal.org/echemportal/substancesearch/page.action?pageID=0 The OECD eChemPortal allows simultaneous searching of reports and datasets by chemical name and number and by chemical property. Direct links to collections of chemical hazard and risk information prepared for government chemical review programs at national, regional and international levels are obtained. Classification results according to national/regional hazard classification schemes or to the Globally Harmonized System of Classification and Labelling of Chemicals (GHS) are provided when available. The list of participating databases can be accessed here: http://www.echemportal.org/echemportal/substancesearch/page.action;jsessionid=1AB4C820B2D854B7FB9381877022B9F6?pageID=2
National Library of Medicine Databases	http://chem.sis.nlm.nih.gov/chemidplus/chemidheavy.jsp Accessed through ChemID Plus, searching on a chemical name or ID produces results that are linked to all NLM databases, including: Registry of Toxic Effects of Chemical Substances (RTECS) ATSDR Public Health Statements ATSDR Toxicological Profiles ATSDR ToxFAQS
TSCATS	The Toxic Substance Control Act Test Submission Database http://www.syrres.com/esc/tscats.htm
California Office of Environmental Health Hazard Assessment	Risk assessment documents prepared by OEHHA on certain Proposition 65 chemicals can be accessed through the links provided in the spreadsheet at: http://oehha.ca.gov/prop65/prop65_list/files/P65list110411links.xlsx
USEPA - Ambient Water Quality Criteria Documents	http://www.epa.gov/waterscience/criteria/wqcriteria.html

Providers/ Data Source	Description
Hazard Information (Data on all toxicological endpoints)	
USEPA – Drinking Water Standards Health Effects Support Documents	http://www.epa.gov/safewater/standards.html
USEPA – ECOTOX Database	http://www.epa.gov/ecotox
IPCS Concise International Chemical Assessment Documents (CICADs)	http://www.inchem.org/pages/cicads.html

APPENDIX B. DATA SOURCES FOR EXPOSURE SCORING

Data Sources for Exposure, Uses, and Environmental Fate (P and B) Scoring

Data Type	Data Source
Uses	Inventory Update Reporting and Chemical Data Reporting (IUR/CDR) Premanufacture Notice (PMN) Database (confidential) Design for the Environment chemicals database (confidential) High Production Volume (HPV) Challenge Submissions EPA Hazard Characterizations and Risk Based Prioritizations OECD Screening Information Assessment Profiles and Reports Screening Information Data Sets (SIDS) Documents National Institutes of Health (NIH) Household Product Database NLM Hazardous Substances Data Bank NLM- Hazmap-Occupational exposure to hazardous agents Source Ranking Database Chemical assessments by other governmental organizations Open literature
Environmental releases	Toxics Release Inventory (TRI) National Emission Inventory (NEI) Database U.S. EPA NIH Hazardous Substances Data Bank
General human exposures, including indoor air contaminants	National Report on Human Exposure to Environmental Chemicals (CDC NHANES) Report to the California Legislature Indoor Air Pollution in California. http://www.arb.ca.gov/research/indoor/ab1173/rpt0705.pdf German Environmental Survey- chemicals in indoor air

(Continued)

Data Type	Data Source
	http://www.umweltbundesamt.de/gesundheite/survey/index.htm NLM Hazardous Substances Data Bank Open Literature
Environmental exposures	National Air Quality System (AQS) U.S. EPA National Contaminant Occurrence Database (NCOD) U.S. EPA Current National Recommended Water Quality Criteria U.S. EPA National Water-Quality Assessment Program (USGS NAWQA) EPA Fish Tissue Studies Clean Air Act Hazardous Air Pollutants (HAPs) Clean Water Act Priority Pollutants Superfund Chemical Data Matrix EPA: Targeted National Sewage Sludge Survey Report Groundwater chemicals Desk reference Chemicals in Groundwater Desk reference 2007 EPA Drinking water Chemical contaminant lists New York State Ambient Air monitoring program California Air Resources Board (ambient air) Washington State Background Soil concentration study NLM Hazardous Substances Data Bank Open literature
Environmental Fate (Persistence and Bioaccumulation)	USEPA: HPVIS Hazard Characterizations prepared by EPA on chemicals in the High Production Volume Challenge Program (HPV): http://iaspub.epa.gov/oppthpv/hpv_hc_characterization.get_report?doctype=2 Risk-Based or Hazard-Based Prioritizations prepared by EPA under the Chemical Assessment and Management Program (ChAMP): http://iaspub.epa.gov/oppthpv/existchem_hpv_prioritizations.report Organization for Economic Cooperation and Development (OECD): eChem Portal http://www.echemportal.org/echemportal/substancesearch/page.action?pageID=0 http://www.echemportal.org/echemportal/substancesearch/page.action;jsessionid=1AB4C820B2D854B7FB9381877022B9F6?pageID=2 SRC Environmental Fate Databases http://www.srcinc.com/what-we-do/efdb.aspx National Library of Medicine Hazardous Substances Databank http://toxnet.nlm.nih.gov/cgi-bin/sis/htmlgen?HSDB Japanese National Institute of Technology and Evaluation (NITE). Biodegradation and Bioconcentration of the Existing Chemical Substances under the Chemical Substances Control Law NITE http://www.safe.nite.go.jp/english/kizon/KIZON_start_hazkizon.html

Appendix C. Derivation of Exposure Scores for Use Types and Release Scores for TRI and Non-TRI Chemicals

Criteria I: Use Type

A variety of use information was reviewed to determine whether chemicals were used for consumer, commercial, or industrial purposes. At least two data sources were used to confirm consumer uses. For example, a reported use in EPA's IUR alone was not deemed sufficient to identify a chemical as being in a consumer product. Also note that many chemicals are present in several different product use and functional use categories. All reported uses were considered, and the use with the highest exposure potential informed the prioritization ranking. See Appendix B for additional information on data sources.

Chemicals that were given a rank of three are believed to be present in consumer products and have high potential for exposure due to widespread uses. Chemicals that received a high score have higher potential for exposure due to high likelihood of releases from the product (off-gassing) and high potential for direct contact during application or use based on close proximity. Examples of product criteria that have an increased likelihood of exposure include: products that are not fully cured (chemical reaction is occurring on-site); products that are spray-applied or brush-applied; products that are liquids, gases, or otherwise have the potential to volatilize; products that have the potential to off-gas, degrade, or otherwise emit chemicals over time; and products that have the potential to be incorrectly applied or used also received a rank of three. Some organizations may identify higher exposure potential uses as being dispersive. Examples of product use categories that have this increased likelihood of exposure include: paints and coatings; adhesives, sealants, and elastomers; building materials such as insulation; soaps and detergents; hair care products; water treatment products; floor coverings; automotive care products; and arts, crafts, and hobby materials.

Chemicals that were given a rank of two had moderate exposure. Chemicals that received a moderate score have moderate potential for exposure because they may be present within a chemically stable matrix; have lower or slower likelihood of release from the product, and have more indirect or bystander exposure. There may be increased distance and time between product sources and individual receptors. These chemicals may slowly off-gas or partition to dust over time. Examples of product use categories include:

plastic and rubber products, electronics products, furniture, and foam seating and bedding products.

Chemicals were given a rank of one if at least one commercial use for that chemical was reported in IUR.

Chemicals that were not reported in IUR or were reported in IUR with industrial uses but no commercial or consumer uses were given a rank of zero for the use type criterion of exposure.

Criteria II: General & Environmental Exposure

A variety of data sources were used to compile information on chemicals present within the environment: ambient air, surface water, groundwater, drinking water, soil, indoor environments (air or dust), and chemicals present within biota (humans, fish, animals, or plants). Only a small percentage of all chemicals are actually measured for in various media for reasons such as a lack of adequate sampling and analytical methods and insufficient resources to collect data. Many of the chemicals identified were not able to be ranked for this criterion due to lack of data.

A summary of the number of chemicals identified in different media is provided below. Note that this compilation of chemicals is an initial effort based on readily available and publicly accessible data. It is not a complete or comprehensive assessment of number of chemicals present in any given environmental or biological media. Approximately two-thirds of these chemicals are on the TSCA inventory while the other one-third is not. Refer to Appendix B for additional information on data sources for each media.

Number of Chemicals Reported in Environmental Media

Occurrence of chemicals (by media)	Number of chemicals
Surface water	401
Ground water	407
Ambient air	409
Soil	270
Indoor environments	300
Drinking water	247
Biota	360
Total	1215

Criteria III: Release Score

III. A. Release Scores for TRI Chemicals

The release score for each chemical was determined using the aggregated releases from the TRI data fields listed in the following table. The 2008 TRI database was used for the chemical ranking scheme. A ranking of 3 was assigned for a sum of releases greater than 100,000 lb/yr, a ranking of 2 for a sum of releases greater than 5,000 lb but less than or equal to 100,000 lb/year, and a ranking of 1 for a sum of releases less than 5,000 lb/yr.

III. B. Release Scores for Non-TRI Chemicals

For chemicals not reported to TRI, 2006 IUR data were used to rank chemicals for potential to be released to the environment. The release ranking was derived based on at least three of the following four factors: (1) IUR Production Volume Ranking; (2) IUR Number of Manufacturing, Processing, and Use Sites Ranking; (3) IUR Industrial and Downstream Processing and Use Ranking; and (4) IUR Commercial/Consumer Use Rankings.

Production Volume and Number of Sites Rankings

For the production volume ranking, data from the non-CBI public IUR database were used to rank chemicals using the following cut-offs: greater than or equal to 1,000,000 lb/year for a high ranking of 3; less than 1,000,000 and greater than or equal to 500,000 lb/year for a medium ranking of 2; and less than 500,000 lb/year for a low ranking of 1.

2008 TRI Data Fields for Release Score

TRI Data Field	
Total Fugitive Air Emissions	Wastewater Treatment (Excluding POTWs)
Total Stack Air Emissions	Landfills/Disposal Surface Impoundments
Total Surface Water Discharge	Surface Impoundment
Total Other On-Site Land Releases (Other Landfills)	Other Landfills
Total Land Treatment	Land Treatment
Total Surface Impoundments	Other Land Disposal
Total Other Disposal	Unknown
POTWs - Total Transfers - Metals Only	RCRA Subtitle C Surface Impoundments (M66)
Transfers To POTWs (Non-Metals)	Other Surface Impoundments (M67)
Transfers To POTWs (Metals And Metal	
Compounds)	

The number of industrial sites ranking, data on manufacturing, processing, and use sites in non-CBI public IUR database were used to rank chemicals using the following cut-offs: greater than or equal to 1,000 sites for a high ranking of 3; less than 1,000 and greater than or equal to 99 sites for a medium ranking of 2; and less than 100 sites for a low ranking of 1.

Industrial Processing and Use (IPU) Ranking

For the industrial processing and use ranking, EPA examined the following codes reported under IUR for each chemicals (see the table of sample categories, below): North American Industrial Classification System (NAICS) code, Process or Use code, and the Industrial Function Category. Each 3-code combination was assigned a ranking (high/moderate/low) based on the potential to be released during the industrial processing/use and downstream use. The Agency ranked each 3-code combination using expert judgment, generic scenarios, and past experience with new and existing chemical assessment. The 3-code combination with highest ranking was used as the score for the IPU ranking for the chemical.

The resulting industrial rankings were modified based on whether the chemical was reported as site-limited by all IUR submitters of that chemical or whether industrial uses may have been required to be reported in IUR. Site-limited chemicals were given an IPU Ranking of 1.

Sample of 2006 IUR Industrial Processing and Use Reporting Categories

Industrial Function Categories	Industrial Processing or Use	Small Sample of NAICS
Adsorbents and absorbents	Processing as a reactant	Petrochemical manufacturing
Adhesives and binding agents	Processing incorporation into formulation, mixture or reaction product	Synthetic dye and pigment manufacturing
Aerosol propellants	Processing incorporation into article	Other basic inorganic chemical manufacturing
Agricultural chemicals (non-pesticide)	Processing repackaging	Resin and synthetic rubber manufacturing
Anti-adhesive agents	Use - non-incorporative activities	Fertilizer manufacturing
Bleaching agents		Paint and coating manufacturing
Coloring agents, dyes		Printing ink manufacturing
Coloring agents, pigments		Plastics bottle manufacturing
Corrosion inhibitors and anti-scaling agents		Tire manufacturing
Fillers		Cement manufacturing

Industrial Function Categories	Industrial Processing or Use	Small Sample of NAICS
Fixing agents		Abrasive product manufacturing
Flame retardants		Ferrous metal foundries
Flotation agents		Electric power generation
Fuels		
Functional fluids		
Intermediates		
Lubricants		
Odor agents		
Oxidizing agents		
pH-regulating agents		
Photosensitive chemicals		
Plating agents and metal surface treating agents		
Processing aid, not otherwise listed		
Process regulators, used in vulcanization or polymerization processes		
Process regulators, other than polymerization or vulcanization processes		
Reducing agents		
Solvents (for cleaning or degreasing)		
Solvents (which become part of product formulation or mixture)		
Solvents (for chemical manufacture and processing and are not part of product at greater than one percent by weight)		
Stabilizers		
Surface active agents		
Viscosity adjustors		
Other		

Under the IUR, reporters had an option to indicate if industrial processing and use (IPU) information was not applicable to their chemical; if all reporters of a chemical indicated that the industrial processing and use information was not applicable, EPA assumed there was no such use and assigned a low ranking of 1. For chemicals with an IPU ranking of 1 or 2 that had one or more IPUs reported as "NRO," the rankings were developed based solely on reported IPUs. No ranking was developed for chemicals with all IPUs reported as "NRO." EPA assigned a high ranking of 3 for chemicals with at least one reported IPU code with a high potential for widespread releases.

Commercial Use (C) Release Ranking

For the commercial use ranking, EPA examined each IUR Commercial Use Code reported for the chemicals and assigned a ranking based on their potential to be released during use. For the purpose of this screening exercise, it was assumed that all the "C" use codes in the 2006 IUR included commercial uses. The Agency used past experience in new and existing chemical assessments of similar chemicals and exposure scenarios, coupled with expert judgment, to examine each use to place the chemical in a high, moderate, or low ranking. The use code with the highest ranking was used as the score for the commercial use ranking for the chemical.

The following table lists samples of rankings associated with certain uses. Commercial uses considered likely to result in air and/or water releases were assigned a high ranking score of 3. Uses with low or no potential for releases were given a low score of 1. The rest of the uses were given a score of 2.

2006 IUR Commercial Use Categories

2006 IUR Commercial Use
C01 Adhesives and sealants
C02 Agricultural products (non-pesticide)
C03 Artists' supplies
C04 Automotive care products
C05 Electrical and electronic products
C06 Fabrics, textiles and apparel
C07 Glass and ceramic products
C08 Lawn and garden products (non-pesticide)
C09 Leather products
C10 Lubricants, greases and fuel additives
C11 Metal products
C12 Paints and coatings
C13 Paper products
C14 Photographic supplies
C15 Polishes and sanitation goods
C16 Rubber and plastic products
C17 Soaps and detergents
C18 Transportation products
C19 Wood and wood furniture

Under the IUR, reporters had an option to indicate if commercial/consumer information was not applicable to their chemical. If all reporters of a chemical indicated that the commercial/consumer information was not applicable, EPA assumed there was no commercial use of the chemical, resulting in a low ranking (i.e., score of 1). For chemicals with a ranking of 1 or 2 that had one or more commercial/consumer uses reported as "not readily obtainable" (NRO) or "Others," rankings were developed based solely on the remaining reported uses. No ranking was developed for chemicals with all commercial/consumer uses reported as "NRO" Or "Others." EPA assigned a High ranking of 3 for chemicals with at least one reported C code with a high potential for widespread releases. If multiple uses were reported, EPA referred to the use code that resulted in the highest ranking.

Scoring Releases for Non-TRI Chemicals

The four ranking scores described above – Production Volume (PV), Number of Sites, Industrial Processing and Use (IPU) ranking, and Commercial Use (C) ranking – were added to develop the release score for non-TRI chemicals. When either IPU or C could not be scored, but all the other factors could be scored, the release score was derived based on the remaining three ranking scores. If neither the IPU nor the C codes could be scored, no release score was assigned to the chemical.

When all four sub-scores were available, the possible total score ranged from 4 to 12, and the non-TRI Release scores were ranked as follows:

High (3) = 9 - 12
Moderate (2) = 7 - 8
Low (1) = 4 - 6

When only three out of the four sub-scores were available (if either IPU or C could not be scored), the possible total score ranged from 3 to 9, and the non-TRI Release scores were ranked as follows:

High (3) = 7 - 9
Moderate (2) = 5 - 6
Low (1) = 3 – 4

The Non-TRI Release score for each chemical was added to the other exposure component scores to derive the Total Exposure Score, as described in the body of this paper.

Appendix D. The TSCA Work Plan Chemicals

TSCA Work Plan Chemicals

The TSCA Work Plan Chemicals Methods Document explains the hazard, exposure, and persistence/bioaccumulation criteria, the data sources used, and how chemicals were scored.

Chemical Name	Hazard Criteria Met	Hazard Score	Exposure Criteria Met	Exposure Score	Persistence & Bio-accumulation Criteria Met	Persistence & Bio-accumulation Score	Use	CASRN
2012 Work Plan Chemicals								
Antimony & Antimony Compounds	Possible human carcinogen Developmental and reproductive toxicity Acute and chronic toxicity from inhalation exposures	3	Widely used in consumer products Present in biomonitoring, drinking water, surface water, ambient air and soil High reported releases to the environment	3	High environmental persistence Moderate bioaccumulation potential	3	Consumer Industrial	Category
1,3,4,6,7,8-Hexahydro-4,6,6,7,8,8,-hexamethylcyclopenta [g]-2-benzopyran *(HHCB)*	Developmental toxicity	2	Widely used in consumer products Present in biomonitoring Estimated to have high releases to the environment	3	Moderate environmental persistence Moderate bioaccumulation potential	2	Consumer Dispersive	1222-05-5

Chemical Name	Hazard Criteria Met	Hazard Score	Exposure Criteria Met	Exposure Score	Persistence & Bio-accumulation Criteria Met	Persistence & Bio-accumulation Score	Use	CASRN
Long-chain chlorinated paraffins (C18-20)	Chronic toxicity to target organs including the liver, kidneys and thyroid Aquatic toxicity	2	Used in commercial/industrial products Present in biomonitoring, surface water and soil	2	High environmental persistence High bioaccumulation potential	3	Industrial Dispersive	Category
Medium-chain chlorinated paraffins (C14-17)	Chronic toxicity to target organs including the liver, kidneys and thyroid Aquatic toxicity	2	Used in consumer products Estimated to have high releases to the environment	2	High environmental persistence High bioaccumulation potential	3	Consumer Dispersive Industrial	Category
Methylene chloride	Probable human carcinogen	3	Widely used in consumer products Present in drinking water, indoor environments, ambient air, groundwater and soil High reported releases to the environment	3	Low environmental persistence Low bioaccumulation potential	1	Consumer Industrial	75-09-2
N-Methylpyrrolidone	Reproductive toxicity	3	Widely used in consumer products Present in drinking water and indoor environments High reported releases to the environment	3	Low environmental persistence Low bioaccumulation potential	1	Consumer Industrial	872-50-4

(Continued)

Chemical Name	Hazard Criteria Met	Hazard Score	Exposure Criteria Met	Exposure Score	Persistence & Bio-accumulation Criteria Met	Persistence & Bio-accumulation Score	Use	CASRN
Trichloroethylene *(TCE)*	Probable human carcinogen	3	Widely used in consumer products Present in drinking water, indoor environments, surface water, ambient air, groundwater and soil	3	High environmental persistence Low bioaccumulation potential	2	Consumer Industrial	79-01-6
Additional Work Plan Chemicals (alphabetical order)								
Acetaldehyde	Possible human carcinogen	3	Used in consumer products Present in drinking water, indoor environments, ambient air and groundwater High reported releases to the environment	3	Low environmental persistence Low bioaccumulation potential	1	Consumer Industrial	75-07-0
Acrylonitrile	Probable human carcinogen	3	Widely used in consumer products Present in indoor environments, surface water, ambient air and groundwater High reported releases to the environment	3	Low environmental persistence Low bioaccumulation potential	1	Consumer Dispersive Industrial	107-13-1
tert-Amyl methyl ether	Chronic toxicity Central nervous system effects	2	Widely used in consumer products Present in drinking	3	Moderate environmental persistence	2	Consumer Industrial	994-05-8

Chemical Name	Hazard Criteria Met	Hazard Score	Exposure Criteria Met	Exposure Score	Persistence & Bio-accumulation Criteria Met	Persistence & Bio-accumulation Score	Use	CASRN
	Potential carcinogenicity to specific target organs		water, surface water and ambient air Estimated to have moderate releases to the environment		Low bioaccumulation potential			
Anthra[2,1,9-def:6,5,10-d'e'f'] diisoquinoline - 1,3,8,10(2H,9H)- tetrone (*Pigment Violet 29*)	Aquatic toxicity	3*	Widely used in consumer products Estimated to have moderate releases to the environment	3	High environmental persistence Low bioaccumulation potential	2	Consumer Industrial	81-33-4
Arsenic & Arsenic Compounds	Known human carcinogen Neurotoxicity Central nervous system effects Acute and chronic toxicity from inhalation exposures	3	Widely used in consumer products Present in biomonitoring, drinking water, surface water, ambient air and soil High reported releases to the environment	3	High environmental persistence Low bioaccumulation potential	2	Consumer Industrial	Category
Asbestos & Asbestos-like Fibers	Known human carcinogen Acute and chronic toxicity from inhalation exposures	3	Widely used in consumer products Present in indoor environments	3	High environmental persistence Low bioaccumulation potential	2	Consumer Industrial	Category

(Continued)

Chemical Name	Hazard Criteria Met	Hazard Score	Exposure Criteria Met	Exposure Score	Persistence & Bio-accumulation Criteria Met	Persistence & Bio-accumulation Score	Use	CASRN
Benzenamine	Probable human carcinogen	3	Used in consumer products Present in ambient air, groundwater and soil High reported releases to the environment	3	Low environmental persistence Low bioaccumula-tion potential	1	Consumer Industrial	62-53-3
Benzene	Known human carcinogen	3	Widely used in consumer products Present in biomonitoring, drinking water, indoor environments, surface water, ambient air, groundwater and soil High reported releases to the environment	3	Low environmental persistence Low bioaccumulati on potential	1	Consumer Dispersive Industrial	71-43-2
Benzo[a] pyrene	Known human carcinogen	3	Present in biomonitoring, drinking water, indoor environments, surface water, ambient air, groundwater and soil	2	High environmental persistence Moderate bioaccumula-tion potential	3	Dispersive Industrial	50-32-8
Benzo(a)anthr acene	Probable human carcinogen	3	Present in biomonitoring, indoor environments, surface water, ambient air, groundwater and soil	2	High environmental persistence Moderate bioaccumula-tion potential	3	Dispersive Industrial	56-55-3

Chemical Name	Hazard Criteria Met	Hazard Score	Exposure Criteria Met	Exposure Score	Persistence & Bio-accumulation Criteria Met	Persistence & Bio-accumulation Score	Use	CASRN
1-Bromo-propane	Possible human carcinogen	3	Widely used in consumer products Present in drinking water, indoor environments, surface water, ambient air, groundwater and soil Estimated to have high releases to the environment	3	Low environmental persistence Low bioaccumula-tion potential	1	Consumer Dispersive Industrial	106-94-5
Butanamide, 2,2'-[(3,3'-dichloro[1,1'-biphenyl]-4,4'-diyl)bis(azo)] bis[N-(4-chloro-2,5 -dimethoxyphenyl)-3-oxo- *(Pigment Yellow 83)*	Acute toxicity	2	Used in consumer products Estimated to have high releases to the environment	3	High environmental persistence High bioaccumula-tion potential	3	Consumer Industrial	5567-15-7
Butanamide, 2-[(4-methoxy-2-nitrophenyl) azo]-N-(2-methoxyphenyl)-3-oxo- *(Pigment Yellow 65)*	Aquatic toxicity	3*	Widely used in consumer products Estimated to have high releases to the environment	3	High environmental persistence Low bioaccumula-tion potential	2	Consumer	6528-34-3

(Continued)

Chemical Name	Hazard Criteria Met	Hazard Score	Exposure Criteria Met	Exposure Score	Persistence & Bio-accumulation Criteria Met	Persistence & Bio-accumulation Score	Use	CASRN
4-sec-Butyl-2,6-di-tert-butylphenol	Chronic toxicity	2	Widely used in consumer products Estimated to have moderate releases to the environment	3	Moderate environmental persistence Moderate bioaccumulation potential	2	Consumer Industrial	17540-75-9
Cadmium & Cadmium Compounds	Known human carcinogen Chronic cardiovascular, renal and musculoskeletal effects Acute and chronic toxicity from inhalation exposures	3	Widely used in consumer products Present in biomonitoring, drinking water, surface water, ambient air and soil High reported releases to the environment	3	High environmental persistence Moderate bioaccumulation potential	3	Consumer Industrial	Category
Carbon tetrachloride	Probable human carcinogen	3	Used in commercial/industrial products Present in biomonitoring, drinking water, indoor environments, surface water, ambient air, groundwater and soil High reported releases to the environment	2	High environmental persistence Low bioaccumulation potential	2	Industrial	56-23-5

Chemical Name	Hazard Criteria Met	Hazard Score	Exposure Criteria Met	Exposure Score	Persistence & Bio-accumulation Criteria Met	Persistence & Bio-accumulation Score	Use	CASRN
p-Chloro-o-toluidine	Probable human carcinogen	3	Present in biomonitoring, surface water and soil	2	Moderate environmental persistence Low bioaccumula-tion potential	2	Industrial	95-69-2
Chromium & Chromium Compounds	Known human carcinogen Reproductive toxicity Developmental toxicity Acute and chronic toxicity from inhalation exposures	3	Used in commercial/industrial products Present in ambient air High reported releases to the environment	2	High environmental persistence Moderate bioaccumula-tion potential	3	Industrial	Category
Cobalt & Cobalt Compounds	Cardiovascular and central nervous system effects Acute and chronic toxicity from inhalation exposures	3	Used in consumer products Present in biomonitoring, surface water, ambient air and soil High reported releases to the environment	3	High environmental persistence Moderate bioaccumula-tion potential	3	Industrial	Category
Creosotes	Probable human carcinogen	3	Widely used in consumer products Present in groundwater and soil High reported releases to the environment	3	Moderate environmental persistence Moderate bioaccumula-tion potential	2	Industrial	8001-58-9

(Continued)

Chemical Name	Hazard Criteria Met	Hazard Score	Exposure Criteria Met	Exposure Score	Persistence & Bio-accumulation Criteria Met	Persistence & Bio-accumulation Score	Use	CASRN
Cyanide Compounds (Limited to dissociable compounds)	Neurotoxicity Reproductive toxicity Central nervous system effects	3	Widely used in consumer products Present in drinking water, surface water and soil High reported releases to the environment	3	Moderate environmental persistence Low bioaccumula-tion potential	2	Consumer Industrial	Category
Dibenz(a,h)anthracene	Probable human carcinogen	3	Present in indoor environments, surface water, ambient air, groundwater and soil	2	Moderate environmental persistence Moderate bioaccumula-tion potential	2	Dispersive	53-70-3
Dibromochloromethane	Possible human carcinogen	3	Present in biomonitoring, surface water, ambient air and soil	2	Moderate environmental persistence Low bioaccumula-tion potential	2	Industrial	124-48-1
3,3'-Dichlorobenzidine dihydrochloride	Probable human carcinogen	3	Used in consumer products Relatively small reported releases to the environment	2	Moderate environmental persistence Low bioaccumula-tion potential	2	Consumer Industrial	612-83-9

Chemical Name	Hazard Criteria Met	Hazard Score	Exposure Criteria Met	Exposure Score	Persistence & Bio-accumulation Criteria Met	Persistence & Bio-accumulation Score	Use	CASRN
1,1-Dichloro-ethane	Mutagenicity	2	Used in consumer products Present in biomonitoring, drinking water, surface water, ambient air, groundwater and soil Moderate reported releases to the environment	3	Moderate environmental persistence Low bioaccumula-tion potential	2	Consumer Industrial	75-34-3
1,2-Dichloro-ethane	Possible human carcinogen	3	Used in commercial/ industrial products Present in biomonitoring, drinking water, indoor environments, surface water, ambient air, groundwater and soil High reported releases to the environment	2	Moderate environmental persistence Low bioaccumula-tion potential	2	Consumer Industrial	107-06-2
1,2-Dichloro-propane	Acute mammalian toxicity	2	Used in consumer products Present in biomonitoring, drinking water, indoor environments, surface water, ambient air, groundwater and soil High reported releases to the environment	3	High environmental persistence Low bioaccumula-tion potential	2	Consumer Industrial	78-87-5

(Continued)

Chemical Name	Hazard Criteria Met	Hazard Score	Exposure Criteria Met	Exposure Score	Persistence & Bio-accumulation Criteria Met	Persistence & Bio-accumulation Score	Use	CASRN
trans-1,2-Dichloro-ethylene	Chronic toxicity	2	Widely used in consumer products Present in biomonitoring, drinking water, surface water, ambient air, groundwater and soil	3	Moderate environmental persistence Low bioaccumula-tion potential	2	Consumer Industrial	156-60-5
p-Dichloro-benzene	Possible human carcinogen	3	Widely used in consumer products Present in biomonitoring, drinking water, indoor environments, ambient air, surface water, groundwater and soil Moderate reported releases to the environment	3	Moderate environmental persistence Low bioaccumula-tion potential	2	Consumer Industrial	106-46-7
o-Dichloro-benzene	Chronic toxicity	2	Widely used in consumer products Present in biomonitoring, drinking water, indoor environments, surface water, ambient air and groundwater	3	Moderate environmental persistence Low bioaccumula-tion potential	2	Consumer Dispersive Industrial	95-50-1

Chemical Name	Hazard Criteria Met	Hazard Score	Exposure Criteria Met	Exposure Score	Persistence & Bio-accumulation Criteria Met	Persistence & Bio-accumulation Score	Use	CASRN
			Moderate reported releases to the environment					
Dichloro-acetic acid	Possible human carcinogen	3	Used in consumer products Present in drinking water	3	Low environmental persistence Low bioaccumulation potential	1	Consumer Industrial	79-43-6
1,2-Dimethoxy-ethane (*Monoglyme*)	Reproductive toxicity Developmental toxicity Chronic toxicity	3	Widely used in consumer products Estimated to have high releases to the environment	3	Low environmental persistence Low bioaccumulation potential	1	Consumer Industrial	110-71-4
1,4-Dioxane	Possible human carcinogen	3	Widely used in consumer products Present in groundwater, ambient air and indoor environments High reported releases to the environment	3	Low environmental persistence Low bioaccumulation potential	1	Consumer Dispersive Industrial	123-91-1
Ethanone, 1-(1,2,3,4,5,6,7,8- octahydro-2,3,8,8-tetramethyl-2-naphthalenyl) -	Aquatic toxicity	3*	Widely used in consumer products Estimated to have high releases to the environment	3	Moderate environmental persistence High bioaccumulation potential	3	Consumer Industrial	54464-57-2

(Continued)

Chemical Name	Hazard Criteria Met	Hazard Score	Exposure Criteria Met	Exposure Score	Persistence & Bio-accumulation Criteria Met	Persistence & Bio-accumulation Score	Use	CASRN
Ethanone, 1-(1,2,3,4,5,6,7, 8- octahydro-2,3,5,5-tetramethyl-2-naphthalenyl)-	Aquatic toxicity	3*	Widely used in consumer products Estimated to have high releases to the environment	3	Moderate environmental persistence High bioaccumula-tion potential	3	Consumer Industrial	54464-59-4
Ethanone, 1-(1,2,3,5,6,7,8, 8a-octahydro-2,3,8,8-tetramethyl-2-naphthalenyl)-	Aquatic toxicity	3*	Widely used in consumer products Estimated to have high releases to the environment	3	Moderate environmental persistence Moderate bioaccumula-tion potential	2	Consumer Industrial	68155-66-8
Ethanone, 1-(1,2,3,4,6,7,8, 8a-octahydro-2,3,8,8-tetramethyl-2-naphthalenyl)-	Aquatic toxicity	3*	Widely used in consumer products Estimated to have high releases to the environment	3	Moderate environmental persistence Moderate bioaccumula-tion potential	2	Consumer Industrial	68155-67-9
Ethylbenzene	Possible human carcinogen	3	Used in consumer products Present in biomonitoring, drinking water, indoor	3	Low environmental persistence Low bio-accumulation	1	Consumer Industrial	100-41-4

Chemical Name	Hazard Criteria Met	Hazard Score	Exposure Criteria Met	Exposure Score	Persistence & Bio-accumulation Criteria Met	Persistence & Bio-accumulation Score	Use	CASRN
			environments, surface water, ambient air, groundwater and soil High reported releases to the environment		potential			
Ethylene dibromide	Probable human carcinogen	3	Used in commercial/industrial products Present in drinking water, indoor environments, surface water, ambient air, groundwater and soil Relatively small reported releases to the environment	2	Moderate environmental persistence Low bioaccumula-tion potential	2	Consumer Industrial	106-93-4
bis(2-Ethylhexyl) adipate	Possible human carcinogen	3	Widely used in consumer products Present in drinking water and indoor environments Estimated to have high releases to the environment	3	Low environmental persistence Low bioaccumulati on potential	1	Consumer Industrial	103-23-1
bis(2-Ethylhexyl) -3,4,5,6-tetrabromophthalate (TBPH)	Developmental toxicity Acute and chronic aquatic toxicity	2	Used in consumer products Present in indoor environments Estimated to have moderate releases to the environment	3	Moderate environmental persistence Moderate bioaccumula-tion potential	2	Consumer Industrial	26040-51-7

(Continued)

Chemical Name	Hazard Criteria Met	Hazard Score	Exposure Criteria Met	Exposure Score	Persistence & Bio-accumulation Criteria Met	Persistence & Bio-accumulation Score	Use	CASRN
2-Ethylhexyl-2,3,4,5-tetrabromobenzoate (TBB)	Developmental toxicity Acute and chronic aquatic toxicity	2	Used in consumer products Present in indoor environments and soil	3	Moderate environmental persistence Moderate bioaccumulation potential	2	Consumer Industrial	183658-27-7
Formaldehyde	Known human carcinogen	3	Used in consumer products Present in indoor environments, drinking water, ambient air and groundwater High reported releases to the environment	3	Low environmental persistence Low bioaccumulation potential	1	Consumer Industrial	50-00-0
Hexabromobiphenyl	Possible human carcinogen	3	Used in consumer products Present in ambient air and soil	2	High environmental persistence High bioaccumulation potential	3	Industrial	36355-01-8
Hexachlorobutadiene	Possible human carcinogen	3	Present in indoor environments, surface water, ambient air, groundwater and soil Relatively small reported releases to the environment	1	High environmental persistence High bioaccumulation potential	3	Industrial	87-68-3

Chemical Name	Hazard Criteria Met	Hazard Score	Exposure Criteria Met	Exposure Score	Persistence & Bio-accumulation Criteria Met	Persistence & Bio-accumulation Score	Use	CASRN
Hexachlorocy clohexane	Possible human carcinogen	3	Present in biomonitoring and surface water	2	High environmental persistence Moderate bioaccumula-tion potential	3	Industrial	608-73-1
1-Hexadecanol	Chronic toxicity	2	Widely used in consumer products Present in surface water, ambient air and soil Estimated to have high releases to the environment	3	Low environmental persistence Moderate bioaccumula-tion potential	2	Consumer Dispersive Industrial	36653-82-4
Lead & Lead Compounds	Neurotoxicity Developmental toxicity Reproductive toxicity	3	Widely used in consumer products Present in biomonitoring, drinking water, indoor environments, surface water, ambient air and soil High reported releases to the environment	3	High environmental persistence Moderate bioaccumula-tion potential	3	Consumer Industrial	Category
Mercury & Mercury Compounds	Neurotoxicity Developmental toxicity Chronic nervous system and hepatic effects	3	Widely used in consumer products Present in biomonitoring, drinking water, indoor environments, surface water, ambient air and	3	High environmental persistence Moderate bioaccumula-tion potential	3	Consumer Industrial	Category

(Continued)

Chemical Name	Hazard Criteria Met	Hazard Score	Exposure Criteria Met	Exposure Score	Persistence & Bio-accumulation Criteria Met	Persistence & Bio-accumulation Score	Use	CASRN
			soil High reported releases to the environment					
4,4'-Methylene bis(2-chloroaniline)	Known human carcinogen	3	Widely used in consumer products Present in ambient air Relatively small reported releases to the environment	2	Moderate environmental persistence Low bioaccumula-tion potential	2	Consumer Industrial	101-14-4
Naphthalene	Possible human carcinogen	3	Widely used in consumer products Present in biomonitoring, drinking water, indoor environments, surface water, ambient air, groundwater and soil High reported releases to the environment	3	Low environmental persistence Low bioaccumula-tion potential	1	Consumer Industrial	91-20-3
2-Naphthalenec arboxylic acid, 4- [(4-chloro-5-methyl-2-sulfophenyl) azo]-3-hydroxy-,	Aquatic toxicity	3*	Widely used in consumer products Estimated to have moderate releases to the environment	3	High environmental persistence Low bioaccumula-tion potential	2	Consumer Industrial	17852-99-2

Chemical Name	Hazard Criteria Met	Hazard Score	Exposure Criteria Met	Exposure Score	Persistence & Bio-accumulation Criteria Met	Persistence & Bio-accumulation Score	Use	CASRN
calcium salt (1:1) *(Pigment Red 52)*								
Nickel & Nickel Compounds	Known human carcinogen Acute and chronic toxicity from inhalation exposures	3	Used in consumer products Present in ambient air High reported releases to the environment	2	High environmental persistence Low bioaccumula-tion potential	2	Consumer Industrial	Category
N-Nitrosodi-ethylamine	Probable human carcinogen	3	Present in biomonitoring, surface water, and ambient air, groundwater and soil Relatively small reported releases to the environment	2	Moderate environmental persistence Low bioaccumula-tion potential	2	Industrial	55-18-5
N-Nitrosodi-methylamine	Probable human carcinogen	3	Widely used in consumer products Present in drinking water, surface water, ambient air, groundwater and soil	3	Moderate environmental persistence Low bioaccumula-tion potential	2	Consumer Dispersive Industrial	62-75-9
N-Nitrosodi-phenylamine	Probable human carcinogen	3	Used in consumer products Present in surface water, groundwater and soil Relatively small reported releases to the environment	2	Moderate environmental persistence Low bioaccumula-tion potential	2	Consumer Industrial	86-30-6

(Continued)

Chemical Name	Hazard Criteria Met	Hazard Score	Exposure Criteria Met	Exposure Score	Persistence & Bio-accumulation Criteria Met	Persistence & Bio-accumulation Score	Use	CASRN
Octamethyl-cyclotetra-siloxane	Reproductive toxicity	2	Used in consumer products Present in biomonitoring, drinking water, indoor environments and surface water Estimated to have high releases to the environment	3	Moderate environmental persistence High bioaccumula-tion potential	3	Consumer Dispersive Industrial	556-67-2
4-tert-Octylphenol *4-(1,1,3,3-Tetramethyl-butyl)- phenol*	Aquatic toxicity	3*	Used in consumer products Present in biomonitoring and drinking water Estimated to have moderate releases to the environment	3	High environmental persistence Moderate bioaccumula-tion potential	2	Consumer Industrial	140-66-9
p,p'-Oxybis(ben-zenesulfonyl-hydrazide)	Reproductive toxicity Mutagenicity	3	Used in consumer products Estimated to have moderate releases to the environment	2	Moderate environmental persistence Low bioaccumula-tion potential	2	Consumer	80-51-3
Pentabromo-phenol	Acute toxicity	3	Used in consumer products Present in surface water and soil	2	High environmental persistence Low bioaccumula-tion potential	2	Industrial	608-71-9

Chemical Name	Hazard Criteria Met	Hazard Score	Exposure Criteria Met	Exposure Score	Persistence & Bio-accumulation Criteria Met	Persistence & Bio-accumulation Score	Use	CASRN
Phthalic anhydride	Respiratory sensitizer	3	Widely used in consumer products Present in groundwater and ambient air High reported releases to the environment	3	Low environmental persistence Low bioaccumulation potential	1	Consumer Industrial	85-44-9
Polychlorinated naphthalenes	Acute dermal toxicity Chronic liver effects	1	Widely used in consumer products Present in biomonitoring	3	High environmental persistence High bioaccumulation potential	3	Industrial	Category
Quartz (Respirable forms only)	Probable human carcinogen	3	Widely used in consumer products Present in drinking water Estimated to have high releases to the environment	3	High environmental persistence Low bioaccumulation potential	2	Consumer Industrial	14808-60-7
Styrene	Possible human carcinogen Central nervous system effects	3	Widely used in consumer products Present in biomonitoring, drinking water, indoor environments, surface water, ambient air, groundwater and soil High reported releases to the environment	3	Low environmental persistence Low bioaccumulation potential	1	Consumer Industrial	100-42-5

(Continued)

Chemical Name	Hazard Criteria Met	Hazard Score	Exposure Criteria Met	Exposure Score	Persistence & Bio-accumulation Criteria Met	Persistence & Bio-accumulation Score	Use	CASRN
1,2,4,5-Tetrachloro-benzene	Chronic toxicity	3	Present in ground water and soil	1	Moderate environmental persistence High bioaccumula-tion potential	3	Industrial	95-94-3
Tetrachloroet hylene *(PERC)*	Probable human carcinogen	3	Widely used in consumer products Present in biomonitoring, drinking water, indoor environments, ambient air, groundwater and soil High reported releases to the environment	3	High environmental persistence Low bioaccumula-tion potential	2	Consumer Dispersive Industrial	127-18-4
Tribromo-methane *(Bromoform)*	Probable human carcinogen	3	Used in consumer products Present in biomonitoring, drinking water, surface water, ambient air and groundwater Moderate reported releases to the environment	3	Moderate environmental persistence Low bioaccumula-tion potential	2	Consumer Industrial	75-25-2
1,1,2-Trichloro-ethane	Possible human carcinogen	3	Widely used in consumer products Present in	3	High environmental persistence	2	Consumer Industrial	79-00-5

Chemical Name	Hazard Criteria Met	Hazard Score	Exposure Criteria Met	Exposure Score	Persistence & Bio-accumulation Criteria Met	Persistence & Bio-accumulation Score	Use	CASRN
			biomonitoring, drinking water, surface water, ambient air, and groundwater and soil Moderate reported releases to the environment		Low bioaccumulation potential			
Triglycidyl isocyanurate	Reproductive toxicity Mutagenicity Acute toxicity from inhalation exposures	3	Widely used in consumer products Estimated to have high releases to the environment	3	Moderate environmental persistence Low bioaccumulation potential	1	Consumer Industrial	2451-62-9
2,4,6-Tri-tert-butylphenol	Chronic toxicity and liver effects	2	Widely used in consumer products Present in indoor environments Estimated to have moderate releases to the environment	3	Moderate environmental persistence High bioaccumulation potential	3	Consumer Industrial	732-26-3
Tris(2-chloroethyl) phosphate *(TCEP)*	Mutagenicity Limited evidence of carcinogenicity	2	Widely used in consumer products Present in drinking water and indoor environments Estimated to have moderate releases to the environment	3	Moderate environmental persistence Low bioaccumulation potential	2	Consumer Industrial	115-96-8

(Continued)

Chemical Name	Hazard Criteria Met	Hazard Score	Exposure Criteria Met	Exposure Score	Persistence & Bio-accumulation Criteria Met	Persistence & Bio-accumulation Score	Use	CASRN
Tris(2,3-dibromopropyl) phosphate *(TBP)*	Probable human carcinogen	3	Widely used in consumer products Relatively small reported releases to the environment	2	Moderate environmental persistence Low bioaccumula-tion potential	2	Consumer	126-72-7
Vinyl chloride	Known human carcinogen	3	Used in consumer products Present in drinking water, indoor environments, surface water, ambient air, groundwater and soil High reported releases to the environment	3	Moderate environmental persistence Low bioaccumula-tion potential	2	Consumer Industrial	75-01-4
m-Xylene	Reproductive toxicity Developmental toxicity	3	Widely used in consumer products Present in biomonitoring, drinking water, surface water, ambient air, groundwater and soil High reported releases to the environment	3	Low environmental persistence Low bioaccumula-tion potential	1	Consumer Industrial	108-38-3
o-Xylene	Chronic toxicity	3	Used in consumer products Present in biomonitoring, drinking water, indoor environments, surface	3	Low environmental persistence Low bio-accumulation	1	Consumer Industrial	95-47-6

Chemical Name	Hazard Criteria Met	Hazard Score	Exposure Criteria Met	Exposure Score	Persistence & Bio-accumulation Criteria Met	Persistence & Bio-accumulation Score	Use	CASRN
			water, ambient air, groundwater and soil High reported releases to the environment		potential			
p-Xylene	Reproductive toxicity	3	Widely used in consumer products Present in biomonitoring, drinking water, surface water, ambient air, groundwater and soil High reported releases to the environment	3	Low environmental persistence Low bioaccumula-tion potential	1	Consumer Industrial	106-42-3

Note: An asterisk (*) in the Hazard Score column indicates the score is based solely on aquatic (environmental) toxicity.

INDEX

A

B

E

F

G

M

N

O

P

Q

R

S

Y